Physical Science for the Elementary Teacher

American University Studies

Series XIV
Education

Vol. 31

PETER LANG
New York · San Francisco · Bern
Frankfurt am Main · Paris · London

Paul B. Otto

Physical Science for the Elementary Teacher

PETER LANG
New York · San Francisco · Bern
Frankfurt am Main · Paris · London

Library of Congress Cataloging-in-Publication Data

Otto, Paul B.
Physical science for the elementary teacher /
Paul B. Otto.
p. cm. – (American university studies. Series
XIV, Education ; vol. 31)
Includes bibliographical references.
1. Science – Study and teaching (Elementary) I. Title.
II. Series.
LB1585.084 1991 372.3'5044 – dc20 90-24212
ISSN 0740-4565 CIP
ISBN 0-8204-1455-7

CIP-Titelaufnahme der Deutschen Bibliothek

Otto, Paul B.:
Physical science for the elementary teacher / Paul B.
Otto. – New York; Bern; Frankfurt am Main; Paris :
Lang, 1991
(American university studies : Ser. 14, Education ;
Vol. 31)
ISBN 0-8204-1455-7
NE: American university studies / 14

Printed in the United States of America.

Table of Contents

Preface

We live in a world inextricably dependent upon science and technology. Tremendous achievements in health, communication, industry, agriculture and domestic life have been made possible through science and technology. Today, a country must maintain strong science and technology to survive economically, militarily and socially.

To maintain leadership in science and technology, it is absolutely essential to maintain a strong educational program in science. Basic to science education is a corps of highly-qualified, motivated teachers. This is especially true of teachers in the elementary schools who daily work with children in their formative years.

The type of science education most effective for elementary school students, as well as those at higher levels, is that which closely models the scientific endeavor. A convincing amount of research evidence indicates hands-on science activity as the superior learning experience for children. It is imperative that elementary school teachers be trained to teach effectively, using hands-on activities to develop an interest in science. It has been the observation of the author, through teaching physical science to preservice elementary teachers, that they do become interested in science by being involved in the activities presented in this work.

This textbook is designed to aid preservice elementary school teachers in achieving competence in teaching science and developing scientific literacy in students. Wherever feasible, activities are recommended as the medium for developing the concepts rather than lecture. It is the philosophy of the author that people learn better by doing than by exhortation. A special feature of this textbook is the inclusion in each chapter of a section on science/technology/society interactions. Many societal problems are associated with science and technology. It is imperative that teachers develop in their students a keen appreciation for the problems in our society resulting from science and technology as well as coping skills to live in this society. The science/technology/society sections in this textbook can aid preservice elementary school teachers to prepare themselves for this effort.

Chapter 1

Why Study Science?

Science and You

Science is a study of the world in which you live. From a physical science perspective, the world consists of matter and the forces which interact with matter.

You may believe that studying forces and the interaction of matter is rather dull, and like most people just beginning to study a subject, you want to know how it relates to your life. Because you will be involved in teaching physical science to elementary school children, you should be interested in learning as much as you can about your physical world. But try to look at it from a strictly personal standpoint as well. Do you need a basic knowledge of physical science to survive in this day and age? Even if you could survive without this knowledge, would your lack of knowledge have negative implications for society in general? The following discussion should help you analyze these questions.

First, consider whether you actually need science and its related area, technology, to survive in the present age. Imagine trying to get along without science and technology. You may feel, for example, that an automobile is a technological luxury. This may be true if you live within walking distance to a shopping center, medical clinic and your job. In most population centers, however, commercial districts are quite removed from residential areas. Of course, you could ride a bus or use some other form of public transportation, but that would merely be substituting one form of technology for another. You could survive without a radio, television or telephone, but you would probably become a dull, uninformed person, who would have difficulty relating with people and holding a job. Even reading a newspaper would be out, as its production requires technology.

Electricity could be eliminated from your home, but just think of all of the uses you have for it. Surviving without it would be very difficult in most residential areas.

You may try to escape technology by moving to a very remote location where you could "live off the land." In such a place you could avoid all modern mechanized conveniences. In fact, you could live in a cave with no telephone, television or radio, electricity or automobile. You could enjoy nature first-hand without concern about the fast-paced, high-tech society. There would be no gadgets to repair or dependence on other people. You would only have to worry about yourself.

There would be, however, the question of survival. You could become self-sustaining and raise your own fruits, vegetables and meat, perhaps supplementing them with wild game and plants. But this type of existence requires an extensive knowledge of the environment. Miscalculating the time for planting or misreading weather conditions could lead to starvation. You would need to be aware of the migration and feeding patterns of wild game to hunt or fish. A lack of knowledge about water drainage patterns might result in your drowning in a cave during a flash flood. "Living off the land" also would require a knowledge of science and technology. You are employing technology when you use any weapon, fishnet, fish hook, tool or fire ignition system, no matter how simple it may seem. Prior to the arrival of Europeans in the Americas, Native Americans had a technology. Although it was devoid of metals, it served their needs quite well. Even a very basic technology requires a knowledge of physical science concepts such as forces, balance, center of gravity, windspeed, and wind direction. The more you know about the characteristic properties of matter and the interaction of forces, energy and matter, the better your chances of improving your technology.

Admittedly, these examples of lifestyles are pretty severe. Few would desire a lifestyle devoid of human interaction and most do not want to do without the modern conveniences made possible by advances in technology. Television, radio, stereos and microwave ovens have become commonplace. Sophisticated technology has become a necessary part of life, and its importance will undoubtedly increase in the furture.

At one time science and technology were considered separate entities. Science was considered to be in the theoretical realm and existed for the sake of knowledge. It was, in fact, referred to as *pure science*. Technology was viewed as the result of applying scientific knowledge to the solution of everyday problems and was referred to as *applied science*.

Modern technology and science are heavily dependent upon one another. Developments in technology cannot take place without the application of the expanding knowledge base of science. Science, in turn, is heavily dependent upon technology to provide the sophisticated apparatuses necessary to conduct scientific investigations. In fact, science and technology have become so interdependent that they are considered almost synonymous.

Suppose you choose not to be a "social recluse," as in the above discussion, but choose instead to be a "science recluse." As such, you would continue to enjoy the fruits of science but would choose to know nothing about science concepts, theories or generalizations; you would not understand the societal implications of science. You would be "scientifically illiterate."

Can you survive today being scientifically illiterate? In this age of high technology, is scientific literacy necessary for the non-science person? Society has passed from the industrial age, to the nuclear age and now into the "information age." Modern high technology deals with information processing, storage and retrieval, and many industrial processes are left to machines such as robots. More and more people are working in service industries. There appears to be very little scientific literacy required for employment in a fast food restaurant, janitorial service or automotive service garage. Of course, all sorts of arguments can be made about the need for workers in service industries to understand heat, solvents and certain principles of mechanical processes. But the hard fact is that most service industries require job-specific training to solve specific problems, most employees in most jobs can survive quite well without knowing the difference between a piston, a camshaft, a resistor or a microchip. If problems arise with a device, a technician is hired to repair it or the whole machine is discarded. Even in your own home, if the refrigerator, stove or dishwasher breaks down or your plumbing becomes stopped, you call a repair person.

The preceding discussion may seem a bit odd for a text about science because it seems to deemphasize the importance of scientific literacy. It certainly does not seem that, beyond the few individuals involved in science or in developing further technology, that most peoples' survival depends on an advanced knowledge of scientific principles. But scientific illiteracy in a large portion of the population does hold serious implications for society as a whole. For example, a person may argue that he/she can waste as much petroleum and electricity as they desire, because, after all, he/she is paying for it. If such an attitude were held by only a few people, it probably would not create problems. But if this short-sighted attitude is shared by a majority of consumers, society as a whole pays the price in terms of greater pollution and resource depletion and higher consumer prices.

No matter what your lifestyle, your world is heavily influenced by science and technology. Mankind has made tremendous strides in the development of science and technology. The accumulation of scientific knowledge and the development of scientific theories has enabled the human race to raise itself from a subsistence level, largely dependent on the whims of nature, to a level where humans can alter and control nature to meet their needs. Advancements in science and technology have led to the construction of skyscrapers, the damming of great rivers and the utilization of heavier-than-air machines for transportation. People can now routinely "fly" in outerspace. Major diseases can be cured; human organs are routinely transplanted to improve the length and quality of life. Life itself can be conceived in a petri dish and genetic patterns altered. It would appear that almost all of the problems of mankind can be solved by science and technology, and there are those who subscribe to such a position.

There are those, however, who would argue that many problems have been created by science and technology, and that each new scientific development causes at least as many problems as it solves. You only need to read a daily newspaper or listen to a news broadcast, to realize that this is true in many instances. Air pollution, water contamination, carcinogens, hypertension, ozone layer alteration, greenhouse effect, over-population and mental anxiety are all examples of societal problems directly related to science and technology.

In a democracy, individual citizens determine the application of technology and the direction it takes at the ballot box. The problem is that individual citizens are often swayed in their decisions merely by promises of economic gains, and they often make voting decisions based on emotion or intuition rather than a solid knowledge base. Because decisions made today can have implications for the future, one of the major goals of education should be to enable the citizenry to be good decision-makers by becoming scientifically literate. A scientifically literate citizen possesses sufficient knowledge of scientific principles, theories, constructs and processes to understand how the physical world functions. Such a knowledge would allow people to look beyond short-term gains to the longer-term implications of their decisions.

As a teacher of science, you can have a strong influence on future voters by helping them become scientifically literate and good decision-makers. It seems reasonable that, in order to become a positive influence, you need to become scientifically literate and possess a good understanding of the societal implications of the use of science and technology.

What is Science?

Science is a process of finding out how the world works. Some say that the term *science* should be a verb rather than a noun because it is an on-going process rather than a fixed set of facts. Facts may be useful for impressing people on game shows or at cocktail parties, but, in and of themselves, they are not the essence of science. Science is an active human endeavor that seeks new information. It is dynamic, not static.

The basis of science is the formulation of a *reasonable theory*. A scientist makes observations and develops an hypothesis based on these observations. They then conduct tests on the object or phenomenon under study by manipulating one *variable* (factors that affect the object or phenomenon) while holding all other variables constant. Theories are formulated through the rejection or support of hypotheses on the basis of the analysis of the results of these tests. Other scientists may then replicate the process in an attempt to disprove the theory. The strength of a theory is determined by its ability to survive scrutiny and testing. A theory that has been accepted for many years may be altered in the face of new discoveries or it may be discarded entirely in favor of a new theory.

Science is said to be *parsimonious* in that it seeks simplicity. If more than one theory is advanced and all explain the phenomenon equally well, the least cumbersome is selected as the most likely. For example, in the epicycle model of planetary motion, each planet was considered to revolve around a circular path (epicycle) centered on the circumference of a very large circle that revolved around the earth, which was placed at the center of the solar system. But the model suggested later by Copernicus placed the sun at the center of the solar system, thus eliminating the need for epicycles. Both theories explained the motion of the planets equally well, but Copernicus' theory was much simpler to represent and is, therefore, the one used today.

Science is not in the business of truth. In actuality, most scientific theories are tentative. A scientific theory does not receive its strength from a position of authority or truth but gains credibility from its ability to predict events. For example, Newtonian mechanics explained the motion of bodies in a particular frame of reference quite well. But at the velocity of light certain discrepancies appeared, and the Einstein theory of relativity was preferred. Even so, the Newtonian model is quite useful for most everyday purposes.

As a student of science you should avoid statements such as: "This proves that such and such is responsible for observed phenomena." Rather, you should develop a habit of using phraseology such as: "The evidence indicates...."; "The most reasonable explanation appears to be...."; or "There appears to be a significant difference between...." Science should not be viewed in terms of absolutes.

It is also important to remember that science is not a religion and, certainly, is not meant as a substitute for religion. Religion is a set of premises considered truths by groups called believers. These truths have usually been pre-established by a divine being and passed on to humankind through some medium. If you accept these premises, then to you they represent facts. These truths will remain unalterable as long as you have faith in them. Because these truths have been received from a divine being, they are not subject to scrutiny. The acceptance of these truths may not be universal. Science is the objective process of explaining observations of nature and is not in competition with religion. Religion has no cause to be threatened by genuine science.

In the following chapters, science will be presented as both knowledge and process. You will develop concepts to explain physical phenomena by doing science and using scientific equipment. Try to read this text from the standpoint of explaining relationships between observed phenomena. Be as concerned about understanding relationships as you are about learning factual material. Both are equally important. One way to obtain a fundamental feel for scientific work is to conduct Experiment 1-1, Educational Destruction.

*****

Experiment 1–1. Educational Destruction

It may seem strange to think of destruction as being educational. When you destroyed something as a child, the only thing you probably learned was not to do it again!

"Educational destruction" could be used to determine the basic materials used to construct a house. Some of the components, such as asphalt shingles, aluminum siding, brick, slate, wallpaper, carpeting and bathroom fixtures, lend themselves to easy analysis because they are so obvious. To find out what is inside the walls (the concrete foundation, the electrical wiring and the plumbing), you would have to take the house apart systematically, piece by piece – in other words, destroy it. Perhaps, the word destroy is a bit too strong. If you take the house apart, you still have the wood, copper, vinyl, concrete, aluminum and other components. Even if you burn the house down, you still have not "destroyed" it. Obviously, the non-combustible elements would be left, but what about the wood? Is something left after burning? What happens during burning? Do you have less material after burning than before? How could you check? What variables would you need to control in order for the experiment to account for everything?

The purpose of this experiment is to determine if wood is more than one substance. In order to find out, you will need to take it apart. What are some ways you could accomplish taking wood apart?

You could examine a piece of wood by looking at it through a magnifying glass. Although this may give you an overall or "global" view of what constitutes wood, it does not give you much insight into what makes up "wood." You could split the piece of wood in half and continue to split it into smaller and smaller pieces until you ended up with sawdust, but this would most likely not add much to your analysis. Observing the sawdust through a magnifying glass would reveal that it appears pretty much the same as the piece of wood with which you began. You would eventually reach a point where you could not make the pieces any smaller. The knife required would need to be too small to handle and you would need a magnifying glass so powerful that the distortions would make it impossible to determine what you were observing.

A second option would be to burn the wood in a closed container in order to keep track of all of the materials (products) that are given off. However, you would need to inject pure oxygen into the container to cause complete burning, which would be a fairly complicated process.

This experiment offers a third choice. Heat will be used to "decompose" wood in a test tube *without burning it*, because air will be driven out by the heat. Any materials given off by the wood will be collected in a test tube and some baby food jars. You can then test these materials and make observations to determine if they are different from each other. It will not be necessary to identify the materials.

Begin the experiment by packing as many wood splints as you can into a large test tube. Insert a one-hole stopper into the test tube with the short leg of a right-angle piece of glass tubing inserted into the stopper as shown in Figure 1.1. Pass the long leg of the glass tubing into a two-hole stopper inserted into a small test tube so that the end of the glass tubing just clears the bottom of the small test tube.

Cool the small test tube in a beaker of cold water. Pass a short-legged piece of right-angled glass tubing just through the two-hole stopper with the opposite end inserted into a piece of rubber tubing .

CAUTION: Always lubricate the glass tubing and stopper hole with glycerine prior to pushing the tubing into the stopper to avoid breakage.

Grasp the tubing close to the end near the stopper hole, thereby reducing its tendency to bend and break while you are pushing. The opposite end of the rubber tubing is placed in a bucket of water in which two inverted baby food jars filled with water rest. Notice the baby food jar lids are placed next to the jars, so once the jars are filled with gas, they can be capped and placed upright on the table for further examination. The open end of the rubber tubing should not be placed under the

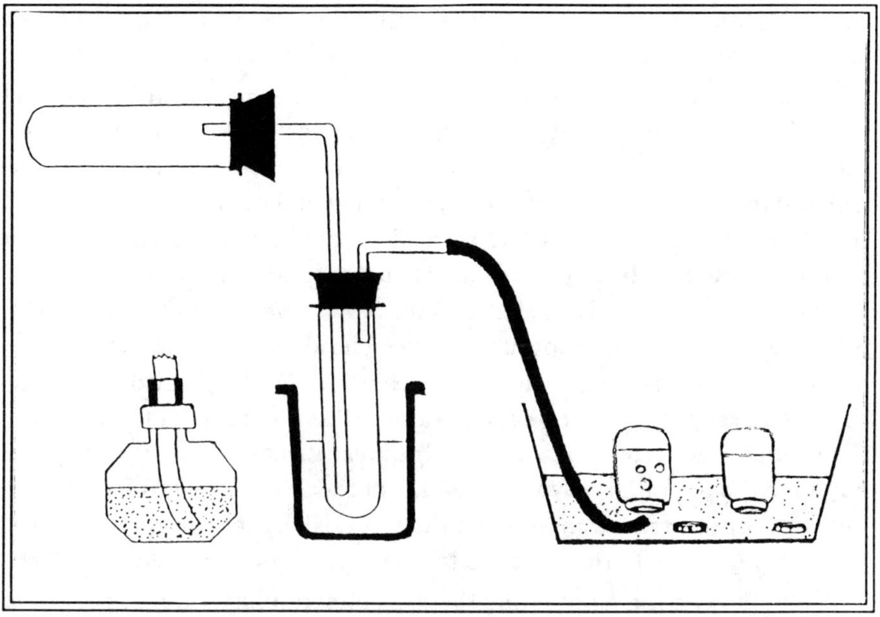

Fig. 1.1. Apparatus for Experiment 1–1.

mouth of either baby food jar at the beginning of the experiment, as initially most of the gas produced will be hot air. After two or three minutes of heating, or as soon as you see "smoke" moving through the glass tubing, you may place the end of the rubber tubing under the mouth of a baby food jar and collect the gas. You will know the jar is full of gas when bubbles move outside of the mouth of the jar. At that time, pull the tube from the jar, slip the cap on the jar and set it on your table top. (Keep the mouths of the baby food jars and the end of the rubber tubing under the surface of the water during the collection procedure.) Next, fill the second jar with the gas. Do not be alarmed if a small amount of water remains in the jars after they have been capped.

When both jars have been filled with gas and capped, take the end of the rubber tubing from the water and lay it on the table top. Remove and extinguish the burner. If you remove the flame prior to removing the

CAUTION: Be sure to place the alcohol burner in a pan of sand for protection in the event of spillage.

rubber tubing from the water, cold water may be forced into the hot, large test tube by atmospheric pressure, causing breakage.

Now, one of the most important parts of the experiment begins—the analysis of the results of the manipulation. Remember, the experiment was set up so that the wood was heated entirely in an enclosed container that isolated the wood from the atmosphere so it would not burn. Any products resulting from heating the wood were channeled through a specific route such that you could account for all of the products. The temperature of the wood is what scientists refer to as the *independent variable*. It was what was manipulated in the experiment. Considerable effort was expended to keep everything else in the experiment constant. Nothing was done to alter the wood except to change its temperature. The condition of the wood and the products given off make up the *dependent variables*. They are dependent upon the temperature of the wood. Another way to look at it is that the number of products given off from the wood is dependent upon what was done to the wood. In this case the temperature was increased.

Take the two jars containing the gas collected into a darkened room. Light a match, remove the lid from a jar and drop the burning match into the jar. Observe any flame, its color and behavior, and record the data. Repeat the process with the second jar.

Next, examine the material in the small test tube that was cooled in the cold water. Hold the test tube up to the light. Observe any differences in appearance of the material in the test tube. Is the material a liquid, solid or gas? Record your observations and any differences you observe in the material inside the test tube.

You may try to separate the materials in the small test tube. Place three or four marble boiling chips into the test tube and mount a one-hole stopper with a right-angle piece of glass tubing into the test tube and secure the test tube clamp, as in Figure 1.2. Connect a piece of rubber tubing to the right-angled glass tubing and terminate the opposite end into a small test tube cooled in a beaker of cold water. Place a wire gauze stand under the liquid in the test tube and heat the liquid with a alcohol burner. Observe the liquid that remains in the heated test tube. Observe any material that forms in the cooled test tube. Do you see differences? Do you see similarities? Do either of the liquids burn? If so, what is given off during the burning? Record your observations.

Fig. 1.2. Apparatus for separating materials in Experiment 1–1.

After the large test tube has sufficiently cooled, tap the wood splints onto the table top. How do the wood splints appear now? Do they still look like wood? Take one of the darkest splints and try to burn it with a match. Does it burn? Does it continue to glow? Do you observe anything being given off? What does the wood splint look like after burning?

Finally, summarize your observations. Remember, you started out with wood, which was heated. The products were a gas, which burned with a specific color(s), a liquid (or was it more than one liquid?) and the remaining wood, which became charred wood.

You began with wood, which is one substance, and ended up with several substances. For purposes here, there is no need to analyze these substances.

*****

Science/Technology/Society

Science and technology have contributed considerably to improving the lives of people. Through science and technology, mankind has taken itself from an existence in which it occupied most of its time maintaining a food supply, shelter and protection from the severe fluctuations in nature, to one where it is possible to engage in other pursuits. Medical technology and procedures have reduced human suffering and increased longevity.

Technology has made many conveniences available that make life easier. Domestic and international travel are routinely undertaken with ease. A variety of inventions (i.e., air conditioning, television and the telephone), which were initially considered luxuries, are now viewed as necessities. People have become so dependent upon machines that there seems to be an obsession with gadgetry. Many of the devices presently found in homes are a matter of convenience rather than necessity.

Each technological development seems to carry with it an associated problem or set of problems. A major by-product of industry is waste materials such as dioxins, solvents and radioactive materials that may be terribly toxic. It is important that the public be as scientifically literate as possible to deal with waste disposal in a satisfactory manner. Burning would appear to be a viable alternative. But burning does not get rid of the constituents, it merely distributes them into the atmosphere. These materials may then be breathed into the lungs or make contact with the skin of a large population of people. Carbon dioxide and carbon are products associated with most types of burning. Distribution of these materials into the atmosphere changes the amount of radiation from the sun striking the earth, resulting in the so-called *greenhouse effect.* Manufacturing and travel are heavily dependent on heat engines that spew enormous amounts of carbon dioxide, carbon particles, sulfides nitrates, sulfur dioxide and water vapor into the atmosphere.

There are many such examples that could be cited. It is important that you realize that in the experiment "Educational Destruction" you did not really destroy anything. Rather, you redistributed the constituents in the wood. The mass of the materials given off is equal to the mass of the wood with which you started.

Landfills are also used for waste disposal, but the space available is quite limited. Furthermore, soil leeching causes toxic waste materials to percolate into the ground water, a major source of human water supplies. In some instances, toxic materials from former landfills have seeped into the basements of housing developments.

Not only is waste disposal a problem associated with manufacturing and travel through the use of heat engines, but, because energy is required, maintaining an adequate supply of fuel is a constant concern. Humans need to realize that present energy sources are finite and most are unreplaceable, at least into the foreseeable future. Wasteful energy practices by individuals accelerate the rate of energy depletion and increase energy costs for everyone. The unnecessary purchase of gadgets and throw-away items with extremely short longevity not only leads to waste disposal problems, but adds to the energy demands in manufacturing. Many of these items are either partially or totally constructed of plastic because of its relative economy and ease of manufacture. These plastic materials are petroleum derivatives and are not biodegradable, thus draining a valuable energy resource and creating disposal problems.

It is anticipated that science/technology/society problems will continue to grow as advancements in science lead to improved technology. Coupled with drastic population increases and the associated energy demand, the outlook for the future may depend on what is done today.

Many believe that some of the science/technology/society problems are caused by a lack of knowledge of scientific principles. A number of science educators and scientists believe that it is necessary to increase the science literacy of the general population of the world to deal with the problems of industrial societies.

Activities

1. List as many of your activities for one day that depend on science and technology.
2. Write down the benefits from science and technology that you observe in the community in which you presently live. Next, record as many negative effects of science and technology as you can from that community. Contrast the two lists.
3. Is being able to control the sex of a child at conception good or evil? Why?

4. Is the sale of the water of major rivers for use in a coal slurry pipeline good or bad? Why?
5. Should Lake Michigan water be transported through a pipeline and sold to Texans for irrigation of farmland?

Chapter 2

Characteristics of Matter

Volume

If you place your finger in a glass filled with water, the water will overflow. If the glass is only partially filled, placing your finger into the water will cause the water level to rise, but the level will return to normal when you remove your finger. If you slip gently into a swimming pool, the change in water level is scarcely noticeable. However, you will most likely notice slapping sound of the water skimmers; the space your body takes up may even make the pool overflow! The space a body takes up or "occupies" is its *volume.* If one body occupies more space than another, it is said that the first body has more volume than the second body.

The space a body occupies depends to a degree upon the state in which it exists—whether it is a liquid, solid, or gas. Water is a good example because it is so common. One can have a glass full of steam, a glass full of liquid water or a glass filled with ice. In essence, all three glasses are filled with water. If the ice in the glass were completely melted, the resultant water would not quite fill the glass. Similarly, if the glass of steam were cooled until it all became liquid water, only a few drops would result. Thus, for a given quantity of matter, the volume may vary considerably, depending on whether it is in a liquid, solid or gaseous state.

Determining the volume of liquids is not difficult because they take the shape of their containers and fill all of the space. For this reason, manufacturers often prefer to ship substances in the liquid state because their space utilization is high.

Gases also take the shape of their containers, but because they have more energy, they are more difficult to contain. For example, if you fail to keep a container of a gas adequately sealed, before long little of the gas will remain. (A point of clarification: Gasoline is commonly referred to as *gas* in the United States. In physical science the term *gas* refers to the *gaseous state of matter.*)

The volume of a gas changes when the pressure is changed. If, for example, you raise the handle of a basketball air pump, air rushes in and fills the cylinder (Fig. 2.1a). After sealing the end of the hose so no air escapes, you can push down on the handle and compress the air into a much smaller volume (Fig. 2.1b). The original volume has been decreased by increasing the pressure. The volume of a gas is also affected considerably by temperature fluctuations. You can get a good feel for the effect of temperature change on the volume of a gas by inflating a small rubber balloon and tying off the open end. When you heat the balloon with a hair blow-dryer, the balloon expands in size and may even break. This demonstrates that increasing the temperature of a gas increases its volume even if the pressure remains constant.

The volumes of solids are not as subject to temperature and pressure conditions as are the volumes of liquids and gases. Solids do not usually

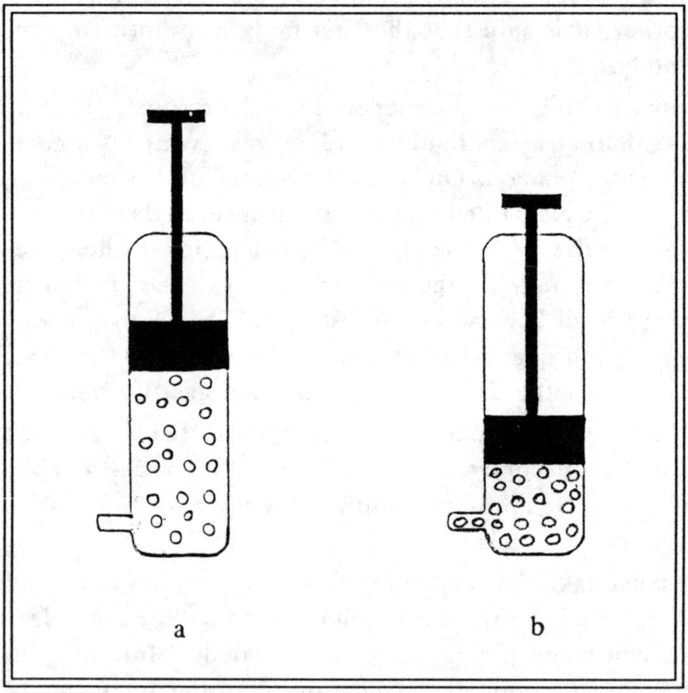

Fig. 2.1. An air pump is a good example of how changing the pressure on a gas changes its volume.

require specialized containers, but solids can present other problems, such as irregular shapes.

The volume of a solid can be determined easily if the single piece being considered has uniform linear dimensions, such as a cube. The volume of a lump of sugar is not found as easily, and determining the volume of a container of several solids is problematic, especially if these solids are not uniform in size, are irregular in shape or both. For example, a greater amount of stone can be placed in a truck if the pieces are small rather than large. More ice can be placed in a water glass if the ice is crushed rather than in cube form. A jar can be completely "filled" with marbles, but a quantity of water can still be poured into the jar without overflowing.

*****

Experiment 2–1. Air In Sand

This experiment will give you practical experience in analyzing the effect of particle size on the overall volume of a solid and allow you to infer practical applications. Carefully measure between 15 and 20 ml (milliliters) of dry sand in a dry 50-ml graduated cylinder and a similar volume of tap water in a second 50-ml cylinder. Accurately record the volumes of each. Carefully pour the dry sand from the graduate into the graduate with the water. Record the total volume of the combined sand and water. Do the combined volumes equal your expected total volume (expected volume equals the arithmetic sums of the two volumes: $V_T = V_1 + V_2$)?

Why is the volume less than the expected value? What caused the decrease in volume? Where did the water go? (Hint: Where did the water go when poured into a jar of marbles?)

The water displaced the air that occupied the space between the individual sand particles. By determining the volume of the water that went between the particles of sand, you can determine the volume of air in the sand. Dividing the amount of air in the sand by the original volume of dry sand yields the *ratio* (fraction) of air in sand. The ratio can be converted into a decimal and stated as a percentage.

*****

It should be clear that determining volume may not be as straight forward as it might seem. Working with the volume of particles on a "macro" scale should help you understand the study of particles on a "micro" scale in subsequent chapters. You may ask: "What's the big deal about the percentage of air in sand anyway?" The fact of the matter is that your life may depend on it! You need water to survive and, most likely, the water you drink comes from a well. Water wells are formed by drilling a hole into the earth until reaching an *aquifer*, which is a layer of sand and gravel with a lot of water between the particles. The greater the surface area of the particles, the more water the layer can hold. A volume of gravel particles, for example, has a greater surface area to which water can cling than does an equal volume of boulders.

The water in an aquifer originated from surface water that percolated through the top soil. In doing so, it carried dissolved minerals along with it, giving the water a distinctive taste. As a comparison, drink some distilled water, which has no dissolved minerals, and notice the bland taste.

Weight

Weight is another concept with which you deal on a daily basis. The problem with weight determination is that it is directly dependent upon the gravitational field in which the weighing is done. An object will weigh more at sea level than it will several thousand feet above sea level because the gravitational pull of the object decreases as the elevation above sea level increases. For everyday activities this difference can be ignored, but if you were to weigh an object on the moon, the weight would be only about one-sixth of its weight on earth. Weight is not a very consistent measure of the quantity of matter when operating in different or varying gravitational fields.

What is needed is a measure of matter that has a common denominator and is consistent throughout the universe. If a quantity of matter is measured in one location, it is essential that the same measurement will result in any other location. You can, for example, appreciate the problem with paying for a pound of candy in one state, when in another state the same amount of candy weighs one-fourth what it did in the first state! *Mass* is a property that is always consistent and is the measure of the amount of matter a body contains. It is independent of gravitational attraction because gravitational factors can be balanced out or the body can be accelerated in a horizontal plane, eliminating gravity as a factor.

Mass

Mass is usually measured in the laboratory on a balance. On the surface this would appear to be weighing the object and therefore subject to the effects of gravity. Gravity does interact with a balance; however, gravity interacts equally with both sides of the balance, so the gravitational effect on one side cancels out the gravitational effect on the other.

Another method of measuring mass that is independent of the effects of gravity is to measure the acceleration of an object moving on a horizontal plane. If the force is kept constant, the greater the mass of the body, the lower the acceleration. You can get a good feel for this by holding an object in your extended arm and swinging it back and forth in a horizontal plane. First, use a ping pong ball and then repeat the procedure holding a large textbook or a clay masonry brick that you are able to hold at arm's length. The greater the mass, the harder it will be to start and stop the object. A hammer that drives a nail a certain distance on earth would be expected to drive the same nail the same distance into the same board if it were located on the moon. The hammer would weigh much less on the moon than on the earth, but the mass would be the same in each frame of reference. The mass of an object is consistent. It does not appear to be dependent on its location or any outside factor as long as it is not traveling at velocities near the speed of light. At or near this point, the mass of a body begins to increase dramatically, but that is beyond the scope of this text.

*****

Experiment 2–2. The Double Pan Balance

The double pan balance (Fig. 2.2) consists of a uniform beam supported by a central pivot point. A pan is supported on each end of the

Fig. 2.2. The double pan balance.

beam. A pointer connected to the beam has its lower end riding above a scale. The double pan balance is a dynamic device in that balancing should take place with the beam swinging back and forth. You should not attempt to get it in balance when it is static (standing at rest) because the friction of the bearings can cause distortions in readings. (This is not true for balances equipped with magnetic dampers.) When balancing masses, the pointer should move the same number of spaces to the left of the center mark on the scale as it does to the right of the center mark.

Prior to doing any mass determinations (technically, you are not weighing) the balance needs to be calibrated – in other words, "zeroed in" or "zero-referenced." The balance would introduce an error in mass determination if it swung further in one direction than the other without any mass placed on the pans.

Calibration begins without any mass on the pans. Riders on each side of the balance allow for fine adjustment. The rider on the right-hand side of the pivot should be set on zero. With the balance moving back and forth, move the left-hand rider until the pointer moves the same number of marks to the left of the center mark on the scale. Balances should be calibrated before each mass determination.

To determine the mass of an object, place the object on the left pan of the balance and standard masses on the right pan of the balance. This is a standard convention that should be followed with any double pan balance you use.

Place a 200-mg mass on the left pan of the balance and move the rider on the right side until *equilibrium* is attained. Mark this point under the rider with a lead pencil. This is the 200-mg mark (0.2 g). Place a mark halfway between this and the zero mark. Half of 0.2 g is, of course, 0.1 g. Now divide the spaces between 0.0 g and 0.1 g, as well as the portion between 0.1 g and 0.2 g, into 5 equal spaces. Each of these spaces is equal to 0.02 g. Now you are ready to do some mass determinations.

First, try to determine the mass of one bead. The obvious approach would be to place one bead on the balance and determine its mass. Due to experimental error, however, you could not place much faith in one observation. A better approach is to determine the mass of a host of beads and average their masses. You can employ a graphical analysis that will reveal a relationship between the mass of the beads and the number of beads. The graph will allow you to directly read the mass of many different beads without placing them on a balance each time. First, construct a table for your data (Fig. 2.3). You can place a zero in the first row of both columns, because zero beads logically would have zero mass.

Count out a number of beads, say ten, and determine their mass. Record the data. Count out another number of beads quite different than the previous number, say in the neighborhood of 25, and determine their masses. To graphically represent your findings, set up a scale on each axis that will cover most of a sheet of paper. Because the mass is a function of the number of beads, plot the number of beads on the horizontal axis (*abscissa*) and mass (*g*) on the vertical axis (*ordinate*). Plot the mass recorded for each of your measurement of beads. Each point plotted is known as a *data point*. Because experimental error is associated with each data point, you should place a small circle around each point to indicate that these points have been determined

# of beads	grams

Fig. 2.3. Table for Experiment 2–2.

experimentally and that you recognize some error may be involved. You would expect two beads to have twice the mass of one bead, three beads three times the mass of one bead, and so on. Therefore, you can reasonably assume that the points will generate a straight line. Lay a straight edge along the *0,0* point and directly through the other two points, if they fall in a straight line. If not, place the straight edge so it passes equidistant from each point (Fig. 2.4) and draw a straight line. This line represents the average of the points and is the line you will use in reading the graph. You should be able to determine the mass of any number of beads by moving up the vertical line until you intersect the plotted curve and then moving horizontally from that point of intersection to read the mass of that number of beads. You can determine the mass of one bead by dividing the vertical reading by the horizontal reading.

*****

Units of Measure

Early mankind probably had little need to determine volume precisely. Terms such as *big, bigger, large* and *larger* were probably accurate enough.

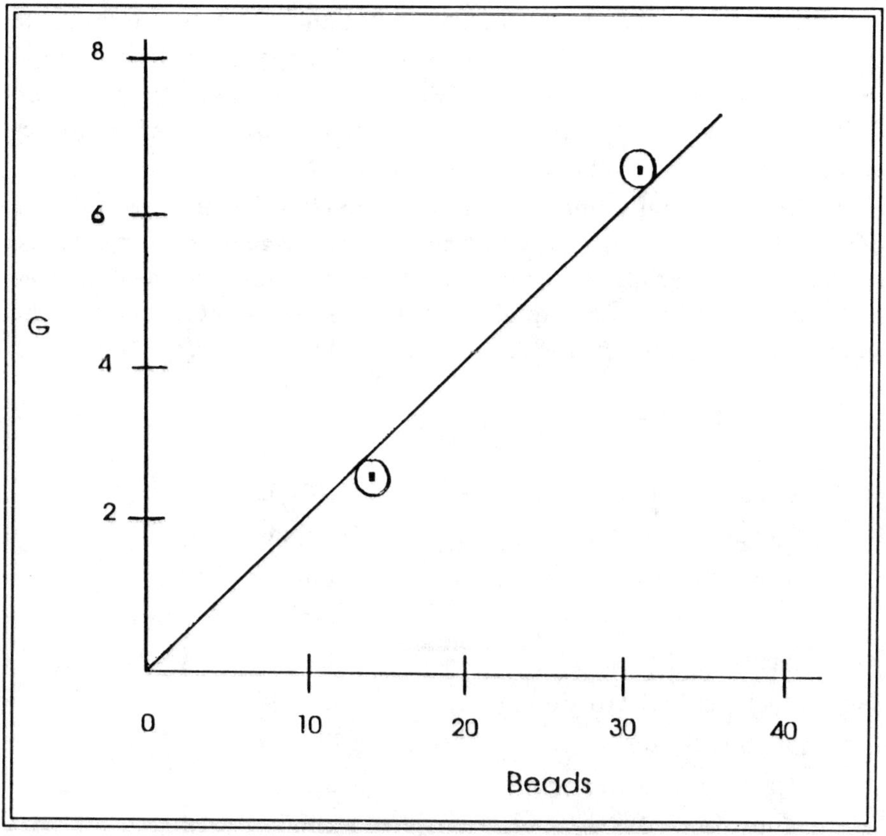

Fig. 2.4. Plotting weights of beads to determine mass.

As societies developed trading systems and began building structures, more exacting standards of measure were required. Arbitrary units of measure such as the cup, quart and gallon were derived from commonly used utensils. Others, such as the cubic foot, were based upon linear measures such as the length of a king's foot. The standard for the inch was the length of a grain of barley. Obviously, kings' feet and grains of barley are not "standard" sizes.

During the French Revolution, because of the desire to change the old to the new, a system of measurement was developed based on multiples of the number 10. Other than in the measurement of time, all units were based upon the circumference of the earth and the density and freezing and boiling temperatures of water. Units were derived such that one could always return to some phenomenon in nature to replicate them. The standard unit of measure was labeled the *meter* (from *metron*, the Greek term "to measure"), which was defined as one ten millionth of the distance from the earth's north pole to the equator. All multiples and subunits were represented by divisions of ten.

Volume, which is a linear measure cubed ($V=L^3$), assumed the unit of the cubic meter (m^3). The cubic meter is too large for most measures, so the *liter* (l) became the standard, being one thousandth of a cubic meter or, in other words, equal to 1,000 cubic centimeters (cm^3). The subdivisions and abbreviations for length and volume are:

Length

1 millimeter (mm)	= 1/1000 meter (m)
1 centimeter (cm)	= 1/100 meter (m)
1 decimeter (dm)	= 1/10 meter (m)

Volume

1000 cubic centimeters (cm^3)	= 1 liter (l)
1000 milliliters (ml)	= 1 liter (l)
1 cubic decimeter (dm^3)	= 1 liter (l)
1000 liters (l)	= 1 kiloliter (kl)

The standard of mass in the metric system of measure is the *gram*. A gram of mass was originally related to linear measure as the mass of one cubic centimeter of water at four degrees Celsius (C), a unit of temperature used in the metric system. Common divisions of mass are:

Mass

1000 milligrams (mg)	= 1 gram (g)
1000 grams (g)	= 1 kilogram (kg)
1 milligram (mg)	= 0.001 gram (g)
1 gram (g)	= 0.001 kilogram (kg)

Temperature

The Celsius temperature scale is used in the metric system. It was named in honor of Anders Celsius (1701–1744), the Swedish astronomer who assigned zero and 100 respectively to the freezing and boiling temperatures of water. He divided the distance between these two temperature points on the thermometer into 100 equal divisions (*degrees*). Thus, the *Celsius temperature scale,* as other units in the metric system, is a function of the base 10. The upper case of the letter *C* is used as an abbreviation of Celsius.

Science/Technology/Society

In the past people had little concern for conserving well water, as it was considered a limitless resource. Recent developments in industrial and agricultural techniques have changed this. Nitrates from fertilizers, both organic and inorganic, insecticides and herbicides have percolated along with surface water into the layers of water-bearing sand into which wells are drilled. Irrigation, industrial processes and the water needs of large population centers have put heavy demands on underground water-bearing strata (*aquifers*) and have drastically lowered the levels of major aquifers. No one really knows how long it would take to return an aquifer to its original level. Some estimate recharging times in terms of hundreds or thousands of years. Some acquifers may be one-shot affairs that will never be replaced by natural process. Because of these problems, considerable conflict has developed over the construction of coal slurry pipelines, irrigation systems and other technologies that require removing large volumes of water from aquifers. This water would not be available then for conventional use by future generations.

That such projects are approved or disapproved by political bodies underlines the need for a scientifically literate citizenry that can make wise political and economic decisions that affect large groups of individuals. If individual voters are knowledgeable about percolation principles, aquifers and ground water levels, they should be equipped to make wise decisions when voting about aquifer-depletion issues. The voting public needs to be constantly informed as to the chemicals that are being approved by government agencies, and their control, in order to head off potential pollution of aquifers.

Farmers and farm managers need to be sensitive to surface run-off patterns and the application of pesticides, herbicides and fertilizers. Contour farming and minimum, or even zero, tillage practices can reduce the amount of surface run-off, reducing the pollution of bodies of water.

Problems

1. A graduated cylinder contains 400 ml of water, all of which is poured into a rectangular container with a bottom 10 cm x 20 cm. What will the depth of the water be in the rectangular container?
2. Notice the top of the liquid in the graduated cylinder in Figure 2.5 makes a curved surface. What is the correct reading of the liquid in the cylinder?
3. What the volume of a piece of copper when immersed in a graduate cylinder containing water compared to a graduate cylinder containing methyl alcohol?

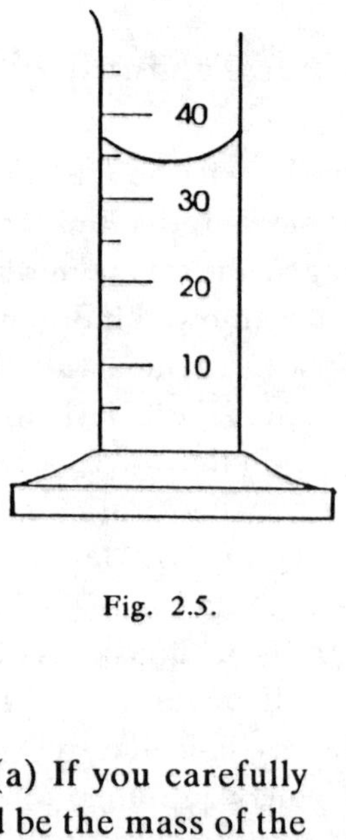

Fig. 2.5.

4. A 10-cm^3 pebble has a mass of 30 g. (a) If you carefully sawed off 1 cm^3 of the stone, what would be the mass of the 1-cm^3 piece? (b) What would be the mass of 60 cm^3 of the pebble?
5. In determining the mass of an unknown object, you place the object on the right-hand pan of a double pan balance instead of the left-hand pan. To balance the unknown mass, a 15-gram mass is placed on the left pan and the rider on the right-hand side is moved to the 0.1-gram mark. What is the mass of the unknown object?

6. A double pan balance is accurate to the nearest 0.02 gram. What would be the expected error in measuring masses of 0.4 grams and 400 grams respectively?
7. What is the volume of a rectangular box measuring 10 cm x 50 cm x 5 cm?
8. When 35 grams of water are added to 60 cm^3 of sand, the total volume is read as 70 cm^3. (a) What volume of water goes in between the particles of sand? (b) What is the volume of the air in the sand? (c) What is the volume of sand only? (d) What percent of the sand is air?
9. What is the density of a cube of metal that measures 5 cm x 10 cm x 5 cm and has a mass of 45 grams?
10. How would you determine the volume of a pebble made of granite? A chunk of sugar that is not rectangular?
11. How would you measure the density of a piece of ice?
12. A 50-ml beaker containing 10 cm^3 of water is placed on the pan of a double pan balance along with an empty 20-cm^3 graduated cylinder. Without spilling, all of the water in the beaker is poured into the graduated cylinder. (a) What will be the volume of water measured in the graduated cylinder, and (b) what will be the mass of the water in the cylinder?
13. A match burns, gives off smoke and a black, charred splinter remains. What would you need to do to show that mass is conserved if you burned another match?
14. Express the following numbers in scientific notation (powers of ten): 10; 1,000; 100,000; 10,000,000.
15. Express the following numbers in scientific notation: 0.001; 0.0005; 0.025; 0.00052.
16. Write the following in base ten: 10^3; 10^7; 10^{10}; 10^6.
17. Express the following in scientific notation: 5,428; 93,000,000; 621; 4,782.
18. Write the following in base ten: 6.35×10^4; 2.21×10^2; 1.3×10^5; 7.82×10^4.

19. Express the following in scientific notation: 0.0025; 0.41; 0.8567; 0.00000092.

20. If a man weighs 180 lbs on earth, approximately what will he weigh on the moon?

Chapter 3

Some Physical Characteristic Properties

Think of two of your friends for a moment. How do you distinguish one from the other? Obviously, their appearance is different! If pressed for specifics, you would probably compare features of the individuals such as height, eye color, hair color, complexion and voice. All of these features are called *characteristic properties* – they allow for the identification of an individual. The use of characteristic properties for identifying substances is also important in the physical sciences. Just as you could not use just one characteristic, say hair color, to identify a friend, you need to use more than one characteristic to identify a substance in the laboratory.

Concepts such as linear measure, volume, mass and weight are not characteristic properties of materials. These concepts apply to specific instances but are not generalizable to the total population. For example, determining that a cube of gold measures one millimeter on each edge does not tell you anything about the properties of the gold. You would be hard-pressed to select that piece of gold from ten seemingly identical cubes made of different alloys of gold. The volume for all the cubes would be the same, so that measurement would not help you discriminate between them. Likewise, weight or mass would not help you make a determination, because both of these measurements depend upon the volume of the material. Such a statement as "iron weighs five pounds" is not logical. You would naturally ask: "How much iron do you have?"

To differentiate materials, you need to define those properties that are unique to the specific substance and, therefore, distinguish it from other substances in *all* instances. In this chapter you will examine several characteristic properties.

Density

Density is the mass per unit volume of a substance. Suppose you are told that one person weighs 220 pounds and another weighs 130 pounds.

Would you judge one person as overweight or the other as underweight based on these measurements alone? It would be difficult, because the first person might be seven feet tall, a veritable string bean, and the second might be five feet tall and plump. Therefore, a more important identifying feature would be each individual's *density*, or weight in relation to the size (volume).

The concept of density is probably used more often than you realize. For example, grains are sold by the bushel, but when farmers sell a truckload of corn, the grain dealer does not measure each bushel. Instead, the corn is weighed and the net weight is divided by the weight per bushel of that grain (*density* = 56 *lbs/bu*) to determine the number of bushels. You have probably experienced the dilemma of trying to determine which package of cold cuts in a supermarket is the best buy—the 12 ounce package for $2.39 or the 16 ounce package for $3.25. What you really need to know is the price per unit or the "price density." Table 3.1 lists some unit prices for four brands of cornflake cereals.

Table 3.1. Prices and unit prices of cereals in a Honolulu, Hawaii, supermarket, 1986.

Brand	mass	price	$/g
General Mills Total Corn Flakes	283	2.81	.0099
Kelloggs Corn Flakes	198	1.25	.0063
Kelloggs Honey Nut Corn Flakes	454	3.21	.0070
Western Family Corn Flakes	680	2.91	.0042

Density can be determined for matter in all three states—solid, liquid or gas. When you compare the densities of two or more substances, however, it is important that the volume of the two substances be precisely the same. For example, if you determine the mass of a piece of styrofoam to be 50 grams (*g*) and a piece of lead to be 10 g, a logical conclusion would be that styrofoam is "heavier" than lead! Intuition would not allow you to accept this, and you would demand that the pieces compared be of equal volume. In other words, you would compare densities. The basic units for density in the scientific world are *grams per cubic centimeter* or the *mass of one cubic centimeter.* It would be perfectly all right to consider density in terms of grams per 10 cubic centimeters or grams per 100 cubic centimeters, but for the sake of simplicity, the volume of one cubic centimeter is used.

Because density is the implied division of grams divided by cubic centimeters, the relationship is represented in mathematical shorthand as:

$$D = g/cm^3 \text{ or } D = \frac{g}{cm^3}$$

Textbooks commonly simplify this further by using one-letter symbols (*formulas*), but that will be avoided here because it is more important to remember relationships than to memorize formulas.

Density can be used to identify substances. Suppose you are to identify the materials that make up five metal rectangular solids. Although all five solids have the appearance of copper, you are aware that some or all may be only copper-coated. You could scratch them to see if they are solid copper, but this would be unsatisfactory to the owner. A density check appears to be your best bet.

Table 3.2 contains the masses, volumes and computed densities for the five solids based upon observations using the same type of balances and rulers you will use in the laboratory. A table of accepted densities reveals the density of cooper as 8.96 g/cm^3. The densities in Table 3.2 appear to be reasonably close to the accepted value for the density of copper, except for sample #4. Are the densities of the samples close enough to the accepted value for copper for you to say with confidence that the specimens are copper? How accurate are your observations?

Table 3.2. Computed densities of five unknown rectangular metallic substances.

Sample	mass-g	volume-cm^3	density-g/cm^3
1	65.52	7.28	9.00
2	90.34	10.06	8.98
3	138.16	15.47	8.93
4	163.46	20.13	8.12
5	61.42	6.87	8.94

One of the major concerns of a researcher is the precision of the instruments used. The calculations of a construct such as density cannot attain greater precision than the least precise of the constituencies of measurement. For example, the balance you calibrated can be read with confidence only to the nearest 0.02 g. As you may recall, that is the smallest division under the rider that can be read with certainty. Therefore, the mass of sample #3, for example, should be reported as 138.16 g ± 0.02 g. The mass reading could be anywhere from 138.18 g (138.16 g + 0.02 g) to 138.14 g (138.16 g – 0.02 g), or ± 0.02 parts in 138.16 g, which can be expressed as ± 0.014%.

You cannot be as precise in volume determination because the best you can do with the ruler used in our experiments is 0.1 mm or 0.01 cm. Therefore, the linear readings cannot be better than ± 0.01 cm. If you assume the edge of one of the cubes measures 3 cm, you would report it as 3.00 cm ± 0.01 cm or 3.01 cm (3.00 cm + 0.01 cm) to 2.99 cm (3.00 cm – 0.01 cm). Cubing this dimension yields a volume of 26.73 cm^3 to 27.27 cm^3 or 27.0 ± 0.27 cm^3, expressed as 27.0 cm ± 1.0% for the ruler. Calculations cannot be expressed with greater precision than ± 1 %.

Which of the densities in Table 3.2 fall within ± 1 % of 8.96 g/cm^3 for copper? For sample #1 the highest density observed experimentally, the 1% range would be from 9.09 g/cm^3 to 8.91 g/cm^3, which includes the accepted value 8.96 g/cm^3. Sample #4, with a density of 8.12 g/cm^3, would have a range of 8.20 g/cm^3 to 8.04 g/cm^3, which would exclude the accepted value of 8.96 g/cm^3. The range of sample #3, density of 8.93 g/cm^3 (the second lowest density), would be from 8.84 g/cm^3 to 9.02 g/cm^3, which includes the accepted value of 8.96 g/cm^3. Similar analysis of samples #2 and #5 reveals that they, too, include the accepted value of 8.96 g/cm^3 for copper. You have good evidence that samples 1, 2, 3 and 5 have the same densities as copper and are likely to be copper. Sample #4, however, has a density significantly different from that of copper and most likely is not copper.

Of course, other factors enter into your analysis. For example, a mistake in measurement could have been made when the data were collected. Before you could be confident of your data, you would want to repeat the measurements and calculations several times. In the experiments conducted in college/university classes, the time available usually allows for only one measurement to be taken. So the data collected by the entire class is analyzed to ensure confidence in your findings.

What assumptions are made with this approach? What variables are introduced during the process?

Notice in the previous discussion that terminology such as the *true value* for the density of copper was avoided. Rather, the term *accepted value* was chosen. This value has been arrived at through a series of measurements and calculations that are agreed upon and certified by the scientific community. As more precise measurements are developed, this value may change. You may notice that some accepted values listed in textbooks change through updated editions. Sometimes errors continue to exist in tables until someone takes the trouble to validate the data. Also, your analysis of Table 3.2 did not lead to the unequivocal conclusion that four samples are copper and one is not. Rather, it was stated that the four samples "appear to have the same densities as copper and are likely to be copper." Rather than stating that "sample #4 is not copper," the statement was made that it "most likely is not copper." Science is not in the business of stating absolutes or "truths." The enterprise attempts to develop theories that can explain or predict phenomena to the best that observations will allow. Try to avoid the trap into which many students fall who, in an attempt to give their speech power, use such terms as *exactly right* or *this proves that*.

*****

Experiment 3–1. Density of a Solid

In this experiment you will attempt to identify the composition of a rectangular solid as either "heavy cube" or "light cube" material. Select one heavy cube, one light cube of the same linear dimensions and a rectangular solid. Determine the volume of each of the three solids using the length of each edge. Find their masses by using the double pan balance. Compute the density of each. Compare the densities of each mass. Do the densities show any similarities? Does the evidence indicate that the rectangular solid is made of "heavy" or "light" material?

Select a pebble and determine its density. Because it has irregular dimensions, how do you determine its volume? Would you be able to use your method with a lump of sugar? Why or why not? Compare the density you found for your pebble with those found by the rest of the class. Is there a significant difference between the densities they found and the one you found? Should there be?

*****

Experiment 3-2. Density of Two Liquids

You may compare two different liquids to see if they are the same or different by looking at them or smelling them.

CAUTION: Never taste any substance in the laboratory. The substance could be poisonous or may react with or dissolve your skin.

A more accurate method of determining whether the liquids are the same is to compare their densities. To do this, you should carefully pour approximately 15 ml of each liquid into separate vials using separate graduated cylinders for each. Will the amount of liquid that "sticks" to each cylinder throw off your readings? Is it the same amount for each liquid? Determine the mass of each volume of liquid. You will, of course, need to subtract the mass of the vials. Should you determine the mass of the vials before or after pouring the liquids into them? Why?

Compute the density for each liquid and compare them. Are they significantly different from each other? If the densities differ, how would you decide if the differences are significant? You may want to average all of the densities found in your class. This may help you decide.

*****

Experiment 3–3. The Density of a Gas

It is much more difficult to determine the density of a gas than that of a liquid or a solid because the volume of a gas is very sensitive to changes in temperature and pressure. Any volume measurement for a gas must be adjusted to "standard" temperature and pressure conditions. Also, some gases are colorless and cannot be seen.

An indirect means will be used to determine the volume and mass of a gas in this experiment. The temperature and atmospheric pressure in the room will be considered to remain constant during the experiment. You may have seen in advertisements or as first-hand observations that some antacid tablets give off a large number of bubbles when placed in water. The bubbles consist of the gas being given off by the tablet. In this experiment, you will collect the gas given off by an antacid tablet in water and determine its mass and volume in order to compute its density.

Break an antacid tablet such as Alka-Seltzer in half and determine the masses of the pieces and a large test tube containing about 10 ml of water as shown in Figure 3.1. (Don't worry if you lose some small pieces, but be sure to find the mass of all the pieces being used in the reaction.) Remember, the mass will be determined for exactly the same items after the reaction of the tablets and water.

Fig. 3.1. Apparatus for Experiment 3–3.

Set up the gas generation and collection apparatus as shown in Figure 3.2. Make sure the free end of the delivery tube extends slightly into the mouth of the jar. Lift the stopper from the large test tube and simultaneously drop all of the Alka-Seltzer pieces into the water inside, noting that the free end of the rubber tubing protrudes into the jar filled with water. Quickly replace the stopper firmly into the test tube and allow the reaction to run for approximately ten minutes.

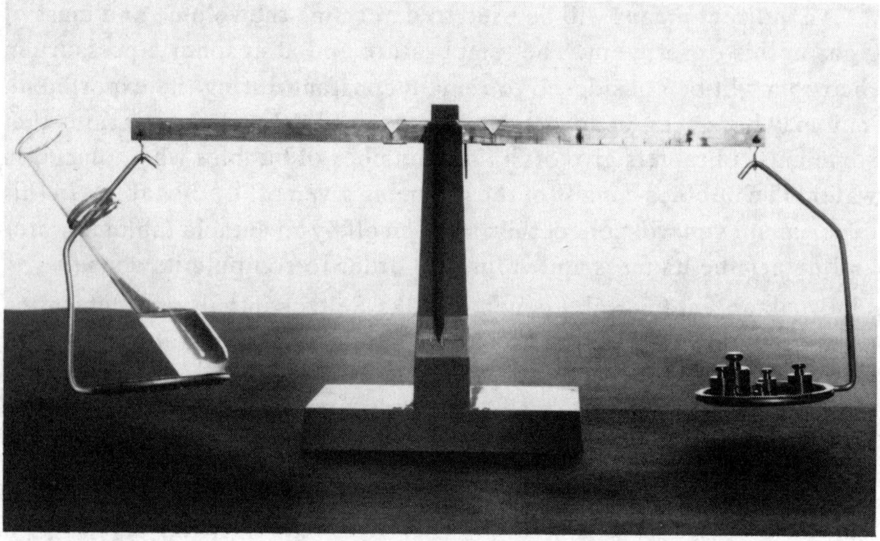

Fig. 3.2. Gas generating and collecting device.

Measure the volume of the water remaining in the jar at the completion of the reaction. What does this volume represent? What represents the volume of the gas? What is the relation of each to the total volume?

Determine the mass of the test tube, water, Alka-Seltzer and rubber band used previously. You would expect a loss in mass from the first reading. Which item actually lost mass? Where does this lost mass appear?

Table 3.3. Densities of some gases at standard temperature and pressure.

Gas	Density g/ml
Air (dry)	1.29×10^{-3}
Ammonia	0.77×10^{-3}
Carbon Dioxide	1.98×10^{-3}
Carbon Monoxide	1.25×10^{-3}
Helium	1.78×10^{-4}
Hydrogen	8.99×10^{-5}
Methane	7.16×10^{-4}
Nitrogen	1.25×10^{-3}
Oxygen	1.42×10^{-3}

Compute the density (mass/unit of volume) for the gas. Determine the average densities found by the class. Is this density similar to the density of any gas found in Table 3.3? A gas with a similar density most likely (although not necessarily) is the same gas. What is the name of the gas you just generated? Can you be certain? (Don't worry if you are not; other tests will be used later to make a more certain determinations.)

*****

Specific Gravity

It is often desirable to compare the densities of various substances. For example, the density of the acid in the lead storage battery used in your automobile can tell you much about its condition. The relationship of the density of a material to the density of a standard, usually water, is named *specific gravity*. Thus, specific gravity has the definition:

$$S.G. = \frac{\textit{density of a substance } (g/cm^3)}{\textit{density of water } (g/cm^3)}$$

In the above relationship the unit g/cm^3 appears in both the numerator and the denominator and can be divided out, leaving specific gravity (*S.G.*) a unitless relationship.

Buoyancy

Buoyancy is the upward force exerted by a liquid upon a body in the liquid. The amount of buoyant force on an object in a liquid is dependent upon the density of the object compared to the density of the liquid. Wood usually floats on water because its density is less than water and, therefore, its S.G. is less than 1.0. As well, a porous stone called *pumice* floats on water because its specific gravity is less than 1.0. A liquid such as motor oil, which is not miscible (does not mix with water) and has a S.G. less than 1.0, also floats on water. One of the problems with petroleum spills on water is that the oil floats on top of the water and sticks to organisms that float on its surface. Buoyancy is really a relative function. For example, the S.G. of iron is about 7.9 and that of mercury, a liquid at room temperature, is 13.5. Therefore, a sphere of iron will float quite high in mercury.

Materials that have an S.G. greater than 1.0 will float on water if their shapes are altered to increase their surface area. They will then displace a mass of water equal to or greater than the mass of the object. The Elementary Science Study (ESS) series includes an activity booklet devoted to making clay boats. By forming modeling clay into shapes ranging from bowls to ships, students learn that clay, which normally sinks, can be made to float by expanding the volume of water it displaces. Likewise, steel, with a S.G. of almost 8.0, will float on water and carry huge cargoes when formed into the shape of a ship. Here again, the mass of the water displaced is balanced with the total mass of the ship and its cargo. Figure 3.3 is an illustration of the buoyancy of a liquid on a floating object. Note that 10 cm^3 of water displaced by the ten gram block (any 10 cm^3 of water) has an equal mass of 10 g ($d = 1\ g/cm^3$).

Although the iron in the block is more dense than water ($S.G. = 7.9$), when shaped into a hollow cube, its outside surface area is increased so that it displaces considerably more water than when it existed as a compact block. Therefore, the iron box will sink until the mass of the water it displaces just equals the total mass of the iron box. If the box were placed in methyl alcohol ($S.G. = 0.79$), a greater volume would need to be displaced to keep the box afloat—in fact, 1.27 times as much. Therefore, the iron box will sink into the alcohol 1.27 times farther than it will sink into water.

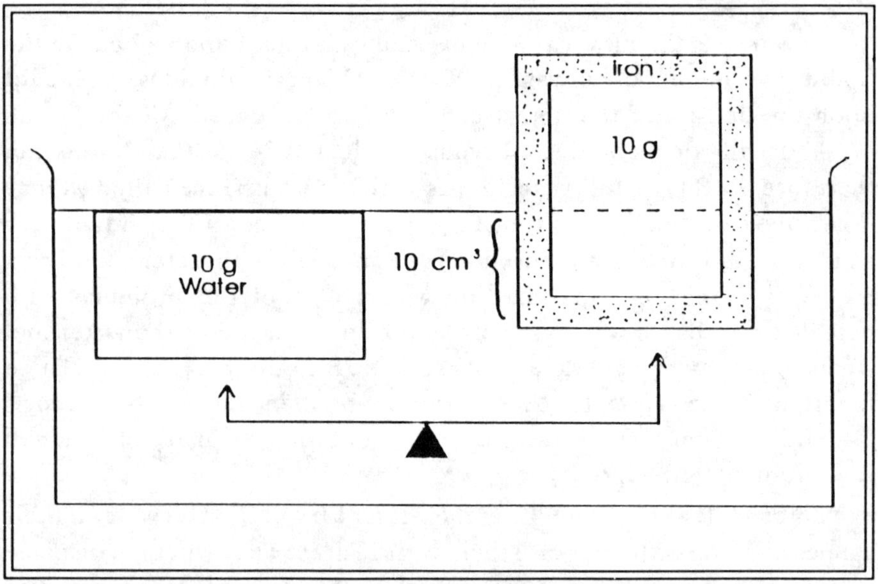

Fig. 3.3 Illustration of buoyancy on water.

Elasticity of Solids

A rubber ball can be squeezed or a rubber band stretched and, when released, seemingly return to their original sizes and shapes. Intuitively, they seem quite elastic. One hardly thinks of a steel wire as being elastic. However, if elasticity is defined as a material's ability to return to its original size, steel may be considered more elastic than rubber!

Is elasticity a characteristic property? Pulling on a steel wire, such as a piano wire, will not reveal any observable change in length. The stretch needs to be amplified so differences can be noted.

The simple apparatus in Figure 3.4 is designed to amplify any change in the length of a wire when a mass is suspended from one end. A wire specimen is fastened to a fixed point, then attached to a pivot on the opposite end and continued beyond the pivot to a bucket. An initial mass of 2.0 kg is placed in the bucket to take all of the kinks out of the wire, and the pivot thumbscrew is adjusted so the long pointer attached to it is "zeroed in" to the lowest mark on the scale. Readings are taken and recorded as successive units of mass are added to the bucket. The wire is removed and the process repeated with wires of two other materials but with the same diameter as the first wire. Inspection of Table 3.4 indicates

Fig. 3.4. Apparatus for measuring elasticity in solids.

that applying the same mass to each wire of equal length and diameter will result in a different change in length. The amount of stretch per unit length appears to be a characteristic property of the material.

Table 3.4. Stretch of different materials of equal length.

		*Wire	
bucket mass	copper	steel	aluminum
2.0	0	0	0
4.0	8	5	13

*wires of equal diameter

Elasticity of Gases -

You can get the feel of the elasticity of a gas by pressing on the handle of a bicycle pump while sealing the outlet. As the force on the handle is increased, the volume of the air in the pump is decreased. The relationship is apparently an *inverse* relationship. In other words, as one factor (*force*) increases, the volume decreases. Conversely, a decrease in force results in an increase in volume.

The change in volume associated with a change in force on a gas can be measured by using the device shown in Figure 3.5. The outlet of the plastic syringe is sealed so the gas is trapped below the rubber piston. A single brick is placed on the plunger of the syringe and the volume is adjusted at some convenient mark. As the number of bricks placed on the plunger increases, the corresponding volume is recorded. The process is conducted with air, Alka-Seltzer gas, and natural gas. Table 3.5 is an example of the relationship of the bricks to the volume for the three gases.

Plotting the data from Table 3.5 yields the graph in Figure 3.6. Notice that the data points for all three gases fall approximately on the same smooth curve, an indication that the elasticity of all three gases is the same.

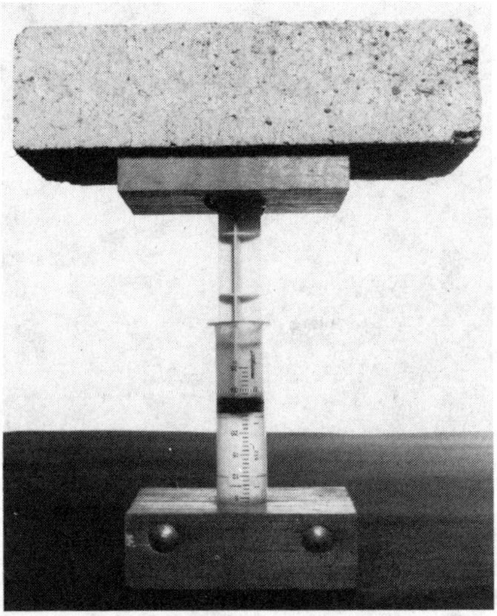

Fig. 3.5. Device for measuring the change in the volume of a gas caused by a change in force.

Also notice that increasing the force (# of bricks) causes the curve to slope smoothly downward. The relationship appears to be an inverse function.

You can test this hypothesis by plotting the inverse of the volume readings from Table 3.6, as in Figure 3.7. Because the volume (V) can be

Table 3.5. Volume as a function of the number of bricks.

	Air	Natural gas	Alka-Seltzer gas
Bricks	Volume (ml)	Volume (ml)	Volume (ml)
1	35.1	35.4	35.3
2	27.7	28.3	28.1
3	23.2	23.1	23.1
4	20.0	20.1	20.0
5	18.2	17.9	18.1
6	16.1	16.5	16.5

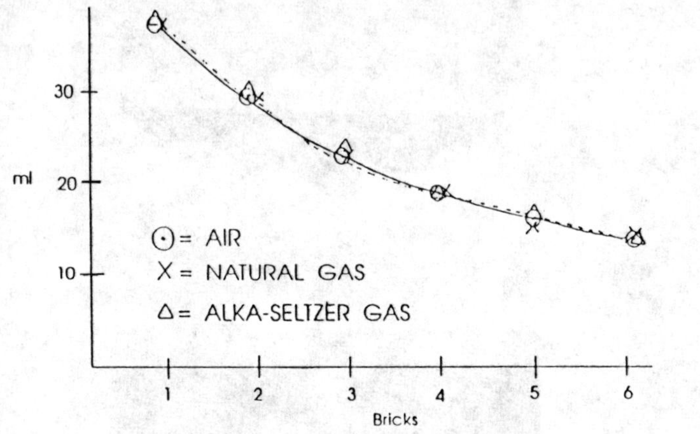

Fig. 3.6. Volume of a gas as a function of applied mass.

thought of as $V/1$, the inverse function is $1/V$. In other words, the inverse values which appear in Table 3.6 are determined by dividing the number one by each of the volume values from Table 3.6. The straight-line plots in Figure 3.7 indicate that the inverse function is the correct relationship.

The slope of the line in Figure 3.7 is the ratio of the vertical axis (*ordinate*) divided by the horizontal axis (*abscissa*), which is the same

Table 3.6. The inverse (reciprocal) of the volume as a function of the number of bricks.

	Air	Natural Gas	Alka-Seltzer Gas
Bricks	Volume $10^{-2} \times cm^{-3}$	Volume $10^{-2} \times cm^{-3}$	Volume $10^{-2} \times cm^{-3}$
1	2.8	2.8	2.8
2	3.6	3.5	3.6
3	4.3	4.3	4.3
4	5.0	4.9	5.0
5	5.4	5.5	5.5
6	6.2	6.0	6.0

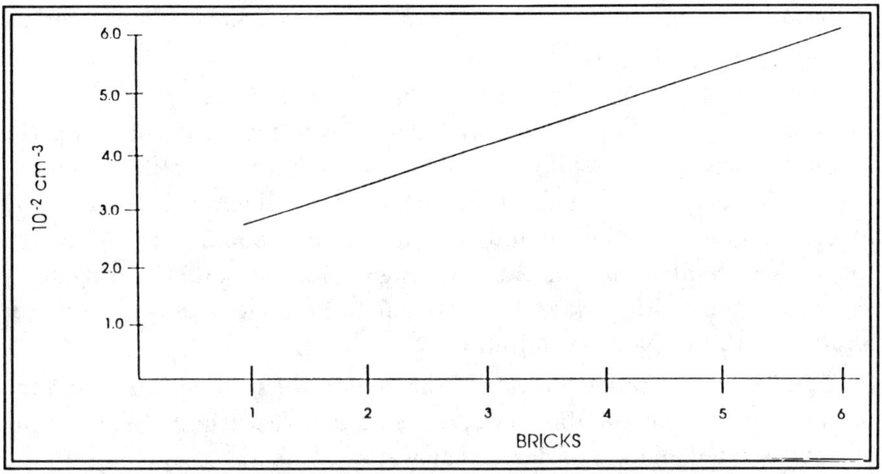

Fig. 3.7. Graphic representation of the data in Table 3.6.

for any point on the line. This ratio is the *coefficient of volume expansion*. The coefficient is the same for the three gases; therefore, it appears the elasticity is the same for all three. In fact, this seems to hold for all gases. Unlike solids, the elasticity of gases is not a characteristic property, as they cannot be distinguished on the basis of elasticity.

Science/Technology/Society

As one approaches Stapleton Airport in Denver, Colorado, it is not unusual to see a brownish-purple haze above the city. It hangs there like a lid on the city and the area up to the beautiful Rocky Mountains. Los Angeles and other cities also have tremendous smog problems, and the smog index report is an every day occurrence. Elderly people and persons with respiratory problems are very sensitive to smog. Epidemiological studies indicate that even healthy people may not fare well in a smog-infested environment over the long haul.

The atmosphere is normally in a dynamic state with an interchange of air masses both horizontally and vertically. When certain atmospheric conditions exist and mankind keeps pumping the debris from automobiles, factories and other forms of combustion, normal interchange may cease and layers of air of different densities may line up vertically, trapping the pollution near the surface of the earth. Terms such as *layer inversion* are used to describe the phenomenon.

There are those who hypothesize that the intense dumping of carbon dioxide into the atmosphere may form a layer that will cut down the amount of sunlight reaching the lower atmosphere, eventually causing another ice age. Others argue that such a layer will allow the short-wave electromagnetic radiation into the lower atmosphere, trapping the longer-length infrared radiation, melting the ice caps and flooding many seacoast cities. Either way, there appear to be serious implications for a future with rampant air pollution.

Density can also play a factor in the ecology of streams, lakes and the oceans. Spraying an oil film on water is an effective method of eradicating mosquitoes. Oil, being less dense than water, floats on top and cuts off the oxygen supply of mosquito larvae. What are the effects of an oil film seal on the ecology of a body of water? How is the food chain affected?

An oil spill from an off-shore drilling platform or an oil tanker can cause all sorts of problems for the environment, both physically and aesthetically. The news media has capitalized on obvious problems such as the detriment to bird life and the spoiling of beautiful beaches with sticky, smelly, black petroleum and the odor of putrifying plant and animal life. As serious as these problems may be, perhaps more significant is the harm being done to marine animals such as oysters, shrimp, lobster and fish. At this point no one really knows the long-range implications of oil spills for the food chain.

What is the natural recovery ability of inland waters as well as the seas? For years mankind had treated bodies of water, such as the Great Lakes and the oceans, as limitless cesspools for industrial wastes. Lake Erie was considered a "dead" lake just a few years ago. Is it not possible that even the oceans have finite limitations? Industry drops entire shiploads of surplus ethylene glycol and other chemical wastes into the Gulf of Mexico. Will there eventually be layers of chemicals of different densities in the oceans such that the marine life and ocean dynamics are permanently altered? Will you live to see the time when lobster, shrimp and oysters are no longer available?

Problems

1. A graph of mass as a function of the number of uniform washers appears in Figure 3.8. (a) Describe a graph of the volume as a function of the number of washers. (b) Describe a graph of the density as a function of the number of washers.

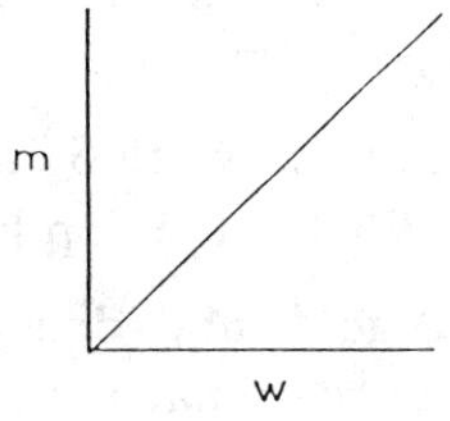

Fig. 3.8.

2. A gas has a volume of 100 cm^3 and a mass of 0.3 grams. (a) What is the density of the gas? (b) If the volume of the gas is changed to 200 cm^3, what is its density?

3. You are given 150 cm^3 of the gas in problem #2a. What is the mass of this volume of gas?

4. The mixing of 100 cm^3 of alcohol (density 0.79 g/cm^3) with 100 cm^3 of water produces a mixture with a density of 0.92 g cm^3. (a) What is the mass of the mixture? (b) What is the volume of the mixture?

5. A rectangular aluminum block measures 6 cm wide x 10 cm high and 20 cm wide. What is its mass (density, 2.7 g/cm^3)?

6. When 750 g of hydrochloric acid reacts with 65.4 g of zinc, a volume of 2240 cm^3 of a gas is generated. If the mass of the solid left in solution after the reaction stops is 20 g, what is the density of the gas?

7. Which of the following characteristic properties (melting temperature, thermal expansion, density, or elasticity) could be used to identify a substance if it is: (a) gas; (b) liquid; or (c) solid?

8. Two pieces of steel wire, each 1 m in length, are fastened to a support and a 1-kg mass is attached to the end of each wire. Wire A has a diameter of 0.1 mm and wire B has a diameter of 0.2 mm. Which wire will stretch more?

9. The mass of sphere A is 77.2 g and its volume is 4 cm^3. Sphere B has a volume of 7 cm^3 and a mass of 79.1 g. (a) Which sphere has a greater density? (b) How much greater?

10. A pan, width 5 cm, length 15 cm, contains 150 g of water. What is the depth of the water?

11. A gold cube (density = 19.3 g/cm^3) measures 0.5 cm on each edge. What is the volume of an equal mass of a cube of crown glass (density = 2.5 g/cm^3)?

12. The mass of 4.5 cm^3 of a gas is 8.1×10^{-3}g. What is its density?

13. Someone tells you that you may have a piece of gold if you can carry it home in your hands. The piece of gold is a cube measuring 20 cm on each edge. Will the gold cube be yours? (density = 19.3 g/cm^3)

14. When a 1-kg brick is placed on a piston of a cylinder containing hydrogen gas, the bottom of the piston comes to rest at the 15-cm^3 mark, and to the 10-cm^3 mark with a 2-kg brick. If the hydrogen gas is replaced with oxygen gas and the exact procedure repeated, what will be the reading of the piston when supporting the 2-kg brick?

15. A wire is 10 cm long when it supports a 4-kg mass. When a 1-kg mass is added to the 4-kg mass, the length of the wire becomes 10.1 cm. What will be the length of a wire 20 cm in length when a 5-kg mass is added to it?

Chapter 4

Some Laws of Motion

In previous chapters you encountered various characteristics of matter as well as buoyant force, which is dependent upon density. Gravitational force was also alluded to, as in the case of the double pan balance. Pressure, which is force per unit of surface area, played a role in the volume of gases. In this chapter you will investigate the nature of forces, how forces interact and the effect of unbalanced forces on matter.

Force

Consider a large, heavy, oak desk standing motionless at the front of a classroom. When you try to move the desk by pushing on it, you observe no motion. If the desk does not move, are you, in fact, exerting a force? When you place a bathroom scale vertically between your hands and the desk, it registers a force, but the desk does not move. The force you exert must be balanced by some other force.

An example of balanced forces producing no motion is the force between the desk and the floor. If you raise the desk and place the bathroom scale under one end, the force the desk exerts registers on the scale. Or, is it the floor exerting a force on the desk? Because no motion is observed, the forces must be equal but in exactly opposite directions.

You can pull a rubber band until it stretches or pull an automobile down a highway. However, you may choose to push the automobile. Forces seem to either be a push or pull.

Inertia

In the case of the desk, if a friend helps you push on the desk, the force is increased to the point where the desk moves to a new location. When you stop pushing, however, the desk stops moving. Once the desk was moving, why did it stop when you stopped pushing? You may shrug it off with: "So it is with desks." Galilei Galileo and Isaac Newton wanted to know more.

During the seventeenth century Gaileo observed that a ball rolling down an inclined plane had a "natural" acceleration and, when rolling up a plane, was decelerated (Fig. 4.1). In a series of "mental" experiments, Galileo developed his idea of constant horizontal motion. When the inclined planes faced each other, the ball would roll down the one plane, across a horizontal surface and up the second incline (Fig. 4.2a). He reasoned that, if friction were eliminated, the ball would roll to the vertical height on the second plane equal to the height from which it was released on the first plane. If the second plane was made less and less steep (Fig. 4ı2b & c), the ball would continue farther and farther horizontally until it reached the release height on the first plane. If the second plane became horizontal, the ball would continue infinitely on the horizontal plane (Fig. 4.2d).

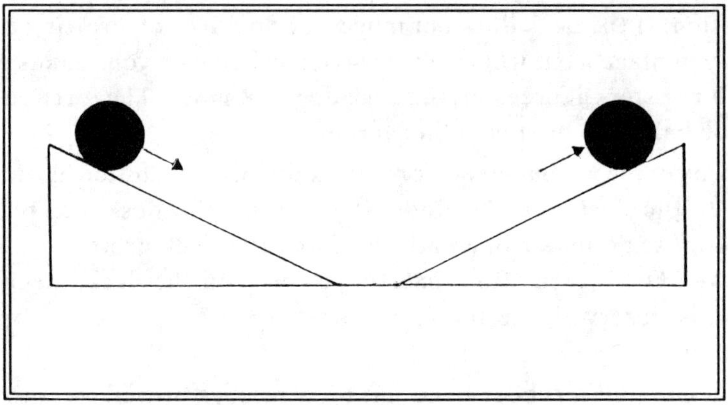

Fig. 4.1. A ball is accelerated down an inclined plane and decelerated up the opposite plane.

Such experiments (perhaps mostly mental) led Galileo to the concept of *inertia*. Essentially, he advocated that a body remains at rest or, once in motion, in uniform straight-line motion unless some outside force acts on it. This concept is referred to as *Galileo's law of inertia*.

If your skeptical nature does not allow you to accept the concept of inertia, you can mentally extend Galileo's reasoning to convince yourself. Take a short run and attempt to slide across a tile floor. You

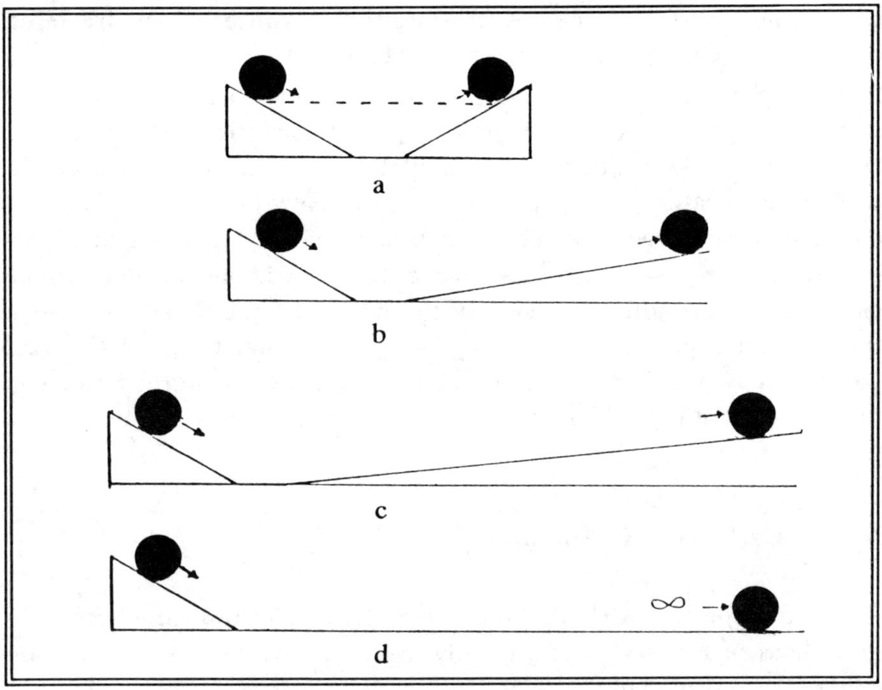

Fig. 4.2. Representations of Galileo's ideas of inertia.

probably didn't slide very far, but note where you stopped. Now, wax the floor and repeat the experiment. Notice that your distance has increased considerably. Imagine that you now use a super wax. This time you smash into the opposite wall! The better the wax you use, the farther you will slide. If you could find a perfectly frictionless wax and an indefinitely long, level floor, you could slide indefinitely and would not stop! Of course, even if you lived indefinitely, you would not be able to check it out, but this type of reasoning can help you conceptualize inertia.

The imaginary wax in your imaginary sliding experiment reduced the opposing conceptual force labeled *friction*. If frictional force is eliminated, bodies will act in an entirely inertial frame of reference. It is, of course, improbable that you could produce the perfect "anti-friction" wax or even a perfectly horizontal surface. Friction can almost be eliminated by the use of gas bearings in the laboratory, resulting in some remarkable demonstrations of inertia. You probably would find such demonstrations to be counter-intuitive, because in your frame of reference, friction always brings bodies to a stop.

Perhaps the closest manifestation of a frictionless frame of reference is the solar system. The observations of planetary motion by Copernicus, Galileo and Tycho Brahe and the calculations of Johannes Kepler based on Galileo's law of intertia allowed Isaac Newton to formulate his *universal law of gravitation*. The planets continuously revolve around the sun, indefinitely it is hoped. No force has been identified that would cause planetary motion, outside of the original force that accelerated them to their present velocities. The gravitational force of the sun continuously accelerates the planets toward itself, thereby causing planetary elliptical orbits rather than continuous straight-line motion through space.

Newton's Laws of Motion

Building on the work of Galileo and others, Newton formulated his basic laws of motion. Traditionally, Newton's works have been summarized into three laws.

Newton's First Law (Galileo's statements on inertia):

> *A body remains at rest or continues in uniform straight-line motion unless influenced by an outside, unbalanced force.*

Actually, you experience inertia daily. When a person rides on the front seat of an automobile without his/her seatbelt fastened, the person and the automobile travel together at the same speed. If the automobile suddenly stops, the passenger without the seatbelt fastened continues to travel and slams into the windshield, often sustaining serious injuries. A belted passenger is stopped by the seatbelt rather than the windshield (Fig. 4.3).

Another common experience with inertia is the removal of a single plastic bag from the perforated roll in supermarket produce departments. Trying to remove a plastic bag by pulling very slowly may produce a length of the roll on the floor. The perforation may be torn to produce one bag

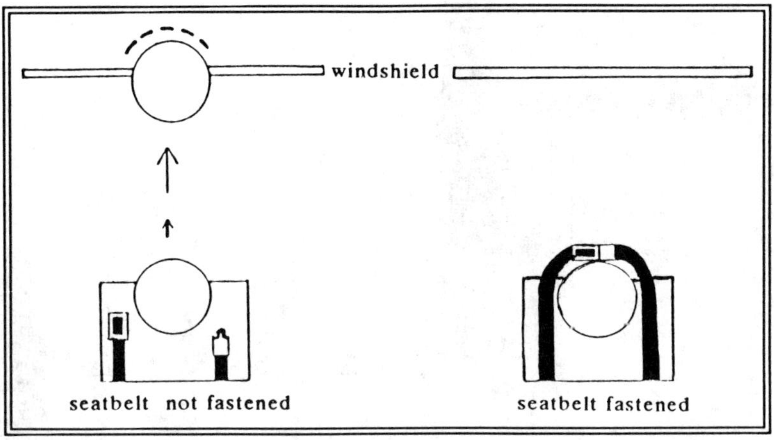

Fig. 4.3. Inertia and the automobile seatbelt.

by giving the end of the bag a vigorous yank (Fig. 4.4). The inertia of the massive roll leaves the roll at rest and the single bag moves forward with a force great enough to tear the perforations to produce an individual bag.

Place a coin on a piece of 3 x 5 cardboard that is resting on a glass tumbler such that the coin is centered over the opening. Holding your forefinger with your thumb, snap your forefinger suddenly against the cardboard. The cardboard flies away, but the coin remains in place horizontally and drops straight down into the tumbler.

Fig. 4.4. Inertia in the produce department.

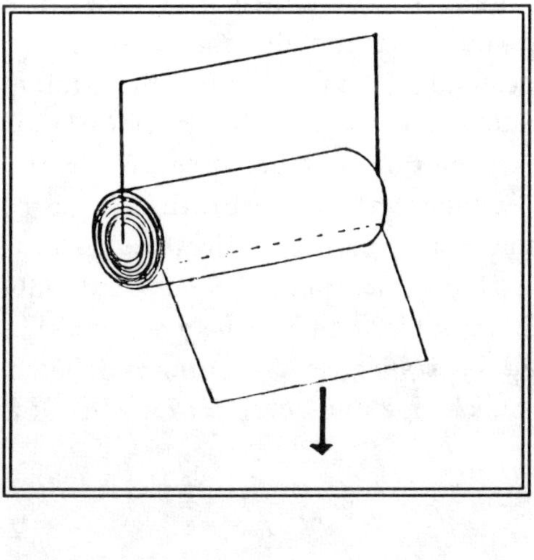

If you are a showperson, you may do a demonstration of inertia for some friends. Place a table-setting on a tablecloth in the center of a small table (Fig. 4.5). Grasp one edge of the tablecloth and give it a quick, brisk yank to remove the cloth. The table-setting will remain in place, undisturbed, due to its inertia. (Failure is due to lack of faith and last-minute hesitation.)

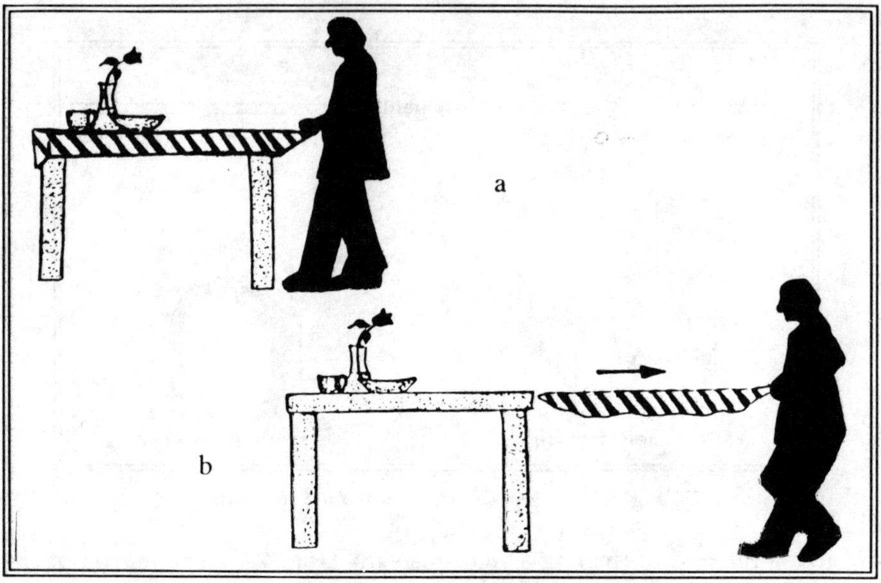

Fig. 4.5. Inertia of a tablecloth. Suddenly jerking the tablecloth away leaves the dishes standing intact.

Newton's Second Law:

> *The acceleration of a body is directly proportional to the applied force on a body and inversely proportional to its mass.*

Prior to the invention of the automobile speedometer, the concept of acceleration was difficult to understand intuitively. When the speedometer reading is changing, you automatically assume the speed is changing. Acceleration can be thought of in the positive or negative direction. If the speedometer reading is increasing, the acceleration is positive; if it is decreasing, the acceleration is negative. When the speedometer reading does not change, the acceleration is zero. Remember, only a changing speedometer reading indicates acceleration. Acceleration is therefore a change in speed. Speed not only changes in magnitude, say from 30 mph to 50 mph, it may also change direction. The concept of speed in a specific direction is labeled *velocity*. Acceleration can also be a change in direction of a body at a constant speed.

The definition of acceleration as a change in velocity is reflected in the units used. Acceleration is expressed in the English System of measurement as miles per hour per second, or mi/hr/sec. This should be interpreted as change in miles per hour for each second. In the metric MKS system (meters-kilograms-seconds), acceleration units appear as m/sec/sec or m/sec^2 and should be considered as changes in meters per second for each second of time considered. A smaller metric unit is the cgs (centimeters-grams-seconds) unit of cm/sec/sec or cm/sec^2.

If you look under the hood of a Ford Escort automobile and then under the hood of a Lincoln Continental, you will notice the Lincoln engine is much larger. Both require about the same acceleration in traffic, so it would seem obvious that the larger auto would require a larger engine than the smaller one. A driver in a drag race wants an automobile with minimum mass and maximum engine force delivered to the driving wheels.

Referring to Figure 4.6, imagine a force applied to a hockey puck on a frictionless surface. Consider the acceleration of the puck with force F_1 applied to it as a_1. If you apply a second force (F_2) of exactly twice the size of the first force, the corresponding acceleration (a_2) is now twice that of using the first force, or $a_2 = 2a_1$. Applying a force three times F_1 yields three times the acceleration. Therefore, you could conclude that acceleration is directly proportional to the applied force. In other words, a change in applied force yields a corresponding proportional change in resultant acceleration. If the force increases, the acceleration increases. A decrease in force results in a proportional decrease in acceleration, when the mass is held constant and all other forces (such as friction) are ignored.

Acceleration can be affected even if the force is kept constant. Figure 4.7 represents an experiment in which the acceleration of one puck with an applied force of F_1 is again a_1. After stacking an identical puck on top of m_1 and again applying the same force (F_1), the acceleration is observed to be just half the acceleration obtained with one puck. In the third case, an identical force (F_1) applied to three ice pucks produces one third the acceleration of one puck. As in the case of applying different forces to the same mass, a proportional relationship exists. The difference is that an increase in mass brings about a corresponding decrease in acceleration

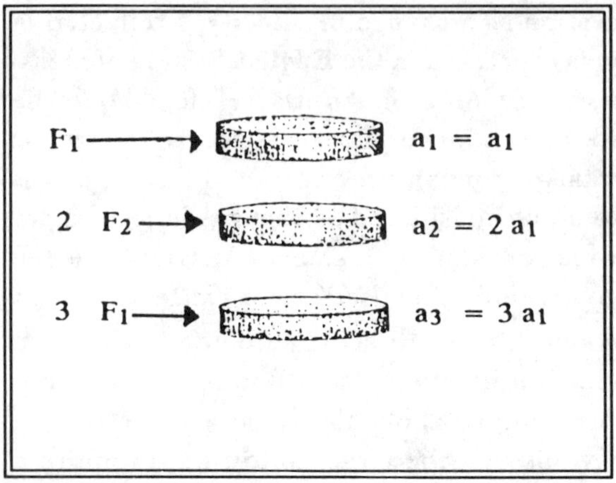

Fig. 4.6. The effect of increasing the force on a constant mass.

for a constant force. This is called an *inverse relationship*. Acceleration is inversely proportional to the mass.

By combining the two observations, you can combine the relationship between force, mass and acceleration and mathematically express Newton's second law as:

$$F = ma$$

Solving for the acceleration yields:

$$a = \frac{F}{m}$$

which really gives a better conceptualization of the two experiments. The force, which is placed in the numerator, indicates that it is "directly" related to the acceleration; whereas, m, being in the denominator, implies an inverse or *reciprocal* relationship.

Newton's Third Law:

> *When a body exerts a force upon a second body, the second body exerts a force of equal magnitude but in the opposite direction upon the first body.*

Suppose an individual stands on a skateboard (Fig. 4.8) and throws a ball as hard as it can be thrown parallel to the wheels. By exerting a

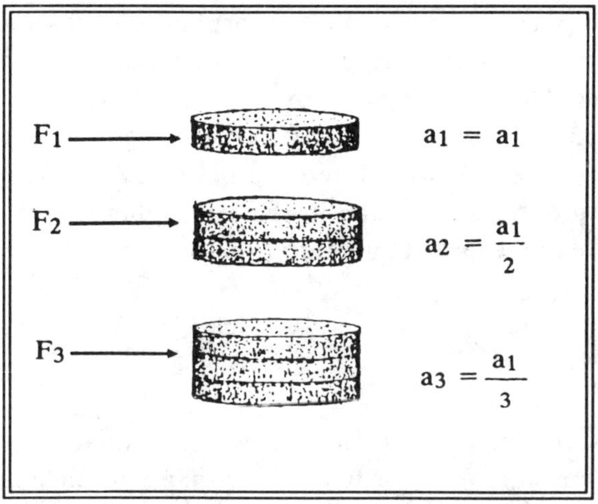

Fig. 4.7. The effect of increasing the mass of a body when the applied force remains constant.

force on the ball, the ball exerts a force in the opposite direction. Newton referred to this as *action/reaction*.

A jet aircraft moves forward because of Newton's third law. The tremendous amount of hot gas being thrown out of the nozzle of the jet engines causes an "action" behind the airplane ($F = ma$) and the opposite "reaction" causes the airplane to travel in the forward direction. The airplane *does not* move forward because the hot gases are

Fig. 4.8. A thrown ball causes the skateboard to accelerate in the opposite direction.

pushing against the atmosphere. Because the atmosphere causes air resistance to the forward motion of the airplane, it would work better in a non-gaseous environment. However, in order to operate in an environment without air, an airplane would need to carry its own oxygen source to burn the fuel, so a rocket engine would need to be used. Rockets operate more efficiently in outer space than in the atmosphere of the earth because in outer space they do not encounter the forward friction of the air.

Impulse/Momentum

Up to this point, the discussion has focused only on the relation of $F = ma$ with little, if any, consideration for time, other than inference in our definition of acceleration. You will recall that acceleration is a change in velocity in respect to a change in time. This can be represented as:

$$a = \frac{\Delta v}{\Delta t}$$

where the Greek letter Δ represents change. Substituting this expression for acceleration into the equation $F = ma$, yields:

$$F = m\frac{\Delta v}{\Delta t}$$

When you multiply both sides of the equation by Δt the equation becomes $F\Delta t = m\Delta v$. The force times the change in time ($F\Delta t$) is labeled *impulse* and mass times the velocity ($m\Delta v$) is labeled *momentum*.

The reason for this devious maneuver, which probably has little meaning to you at this point, was to rewrite Newton's second law of $F = ma$ so that an expression of time appears on one side of the equation and velocity appears on the other side. The formula of $F = ma$ does not directly indicate how long an automobile would need to be pushed to get it to a certain speed. In order to be useful, the relationship must be represented as a force applied for a specific time period that is equal to the mass of a body mutiplied by the final velocity of the body.

Impulse and *momentum* are two relationships that can be used to explain numerous phenomena, even at the subatomic level. Often, many forces need to be considered simultaneously with geometric configuration.

Impulse and momentum can help you understand the "Newton Spheres" devices featured in novelty stores. A number of chrome-plated steel spheres of the same diameter are suspended by strings. Pulling an end sphere away from its rest point and allowing it to strike the next sphere causes little displacement of the inside spheres, but the sphere at the opposite end magically swings out to approximately the same height as you displaced the first sphere (Fig. 4.10). Displacing two spheres, causes two spheres at the opposite end to swing away, etc.

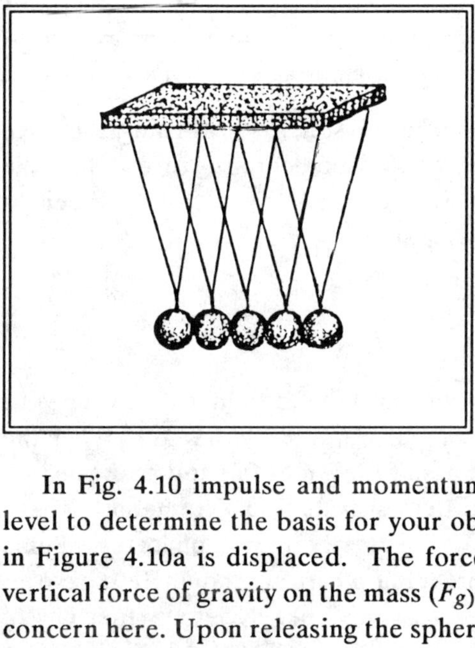

Fig. 4.9. Newton's spheres apparatus.

In Fig. 4.10 impulse and momentum are employed at a rudimentary level to determine the basis for your observations. The left-most sphere in Figure 4.10a is displaced. The force on the string (F_s) is due to the vertical force of gravity on the mass (F_g) and the horizontal force (F_h), the concern here. Upon releasing the sphere, F_h will act for time segment Δt, until the sphere strikes the next sphere. Though F_h becomes smaller as the sphere descends, the mathematics involved in averaging this factor is beyond the scope of this text, so it will be overlooked here. As the sphere strikes the first sphere to the right, it has momentum ($m\Delta v$) due to the impulse ($F\Delta t$). Being a very hard sphere with close-to-perfect elasticity, it transforms its $m\Delta v$ to the second sphere, of like characteristics, which in turn transmits its momentum to the third sphere. The process is duplicated until the momentum is transmitted to the end sphere that has

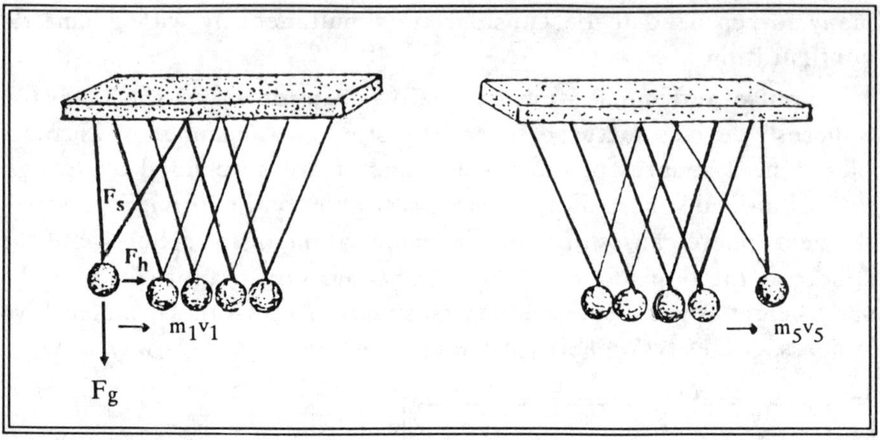

Fig. 4.10. The forces of impulse and momentum acting on a Newton's sphere apparatus.

no place to transmit its momentum. The end sphere rises against the force of gravity and gains potential energy until its momentum is zero at the same height at which the first sphere was released (Fig. 4.10b). Magic is debunked with minimal loss of fascination!

Centripetal Force

Centripetal force is yet another concept related to inertia. Swing your arm vertically while holding a bucket partially filled with water. When the bucket is directly overhead, why doesn't the water fall and soak you? A television automobile commercial demonstrated an automobile driving through a vertical loop (Fig. 4.11). Why doesn't it crash on its top? Some would ascribe it to centrifugal force, but a careful examination exposes "centrifugal" force as a phony force. Remember, Newton's first law of motion provides that a body at rest or in uniform straight-line motion remains so unless acted upon by an outside, unbalanced force.

In Figure 4.12 a body is moving at constant velocity in a straight line. A force is applied periodically at right angles to the body through an attached string. The result is a curved path. If the force on the string becomes constant, such as fixing the end of the string, the direction of the speed of the mass is constantly changed toward the center of the circle. Remember, a change in velocity is acceleration. The acceleration

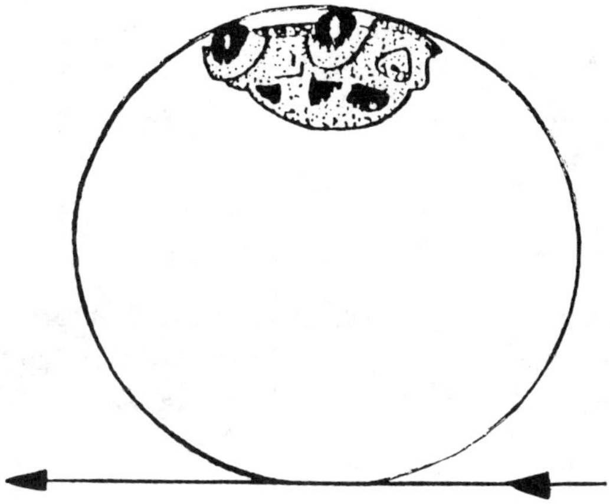

Fig. 4.11. The automobile negotiates the vertical track, apparently defying gravity!

of the body is toward the pivot point and along the radius.

Because the radius is at a right angle to any point on the circle, the acceleration is at a right angle to the velocity. The string causes acceleration inward along the radius and provides the necessary force. This force is called *centripetal* force.

If the string is cut, the mass will again continue in a straight line at a right angle to the radius at the time of release. Because the mass does not move in a direction along the radius opposite that of the centripetal force, no force apparently existed. Such a phony force, labled *centrifugal*

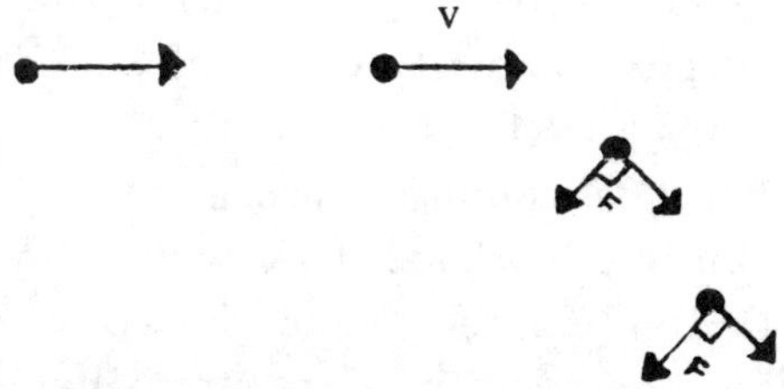

Fig. 4.12. Force applied to a body moving at a constant velocity.

force, is invented to balance the centripetal force when one is in the frame of reference of the sphere in circular motion. Certain liquid pumps are still marketed as "centrifugal pumps." A "centrifuge" is still employed in a chemistry laboratory. The misnomer does not seem to affect the operation of these machines.

*****

Experiment 4–1. Centripetal Force

Set up the apparatus as shown in Figure 4.13. The centripetal force on the whirling stopper is directed along the fishline downward through the glass tubing and is just balanced by the force of gravity on the washers with mass (m).

CAUTION: Make sure the stopper has been securely fastened to the string so as not to injure another person.

The object of the experiment is to determine the relationship of the centripetal force to the mass, speed and radius of the circle. You can get a feel for the relationships by whirling the stopper while holding on to the string below the glass tube with only one washer on the end of the string. As you increase the speed of the stopper, do you have to pull harder on the string? If you let go of the string, what happens to the stopper? What direction does it go?

You will determine a relationship of the period of revolution of the rubber stopper to the centripetal force, where the period T is the time required to make one revolution and the force is represented in units of "washers." Using the same stopper assures a constant mass. The radius is held constant by observing the position of a tiny piece of masking tape in respect to the bottom end of the glass tubing.

Slide enough string through the glass tubing to form a radius of about 100 cm, when the stopper is whirling. Mark the string with a tiny piece of masking tape just below the bottom end of the glass tubing. The marker should in no way inhibit the movement of the string, but should serve only as a marker. Slip three washers over the paper clip on the end of the string. Whirl the stopper in a horizontal plane just so the tape

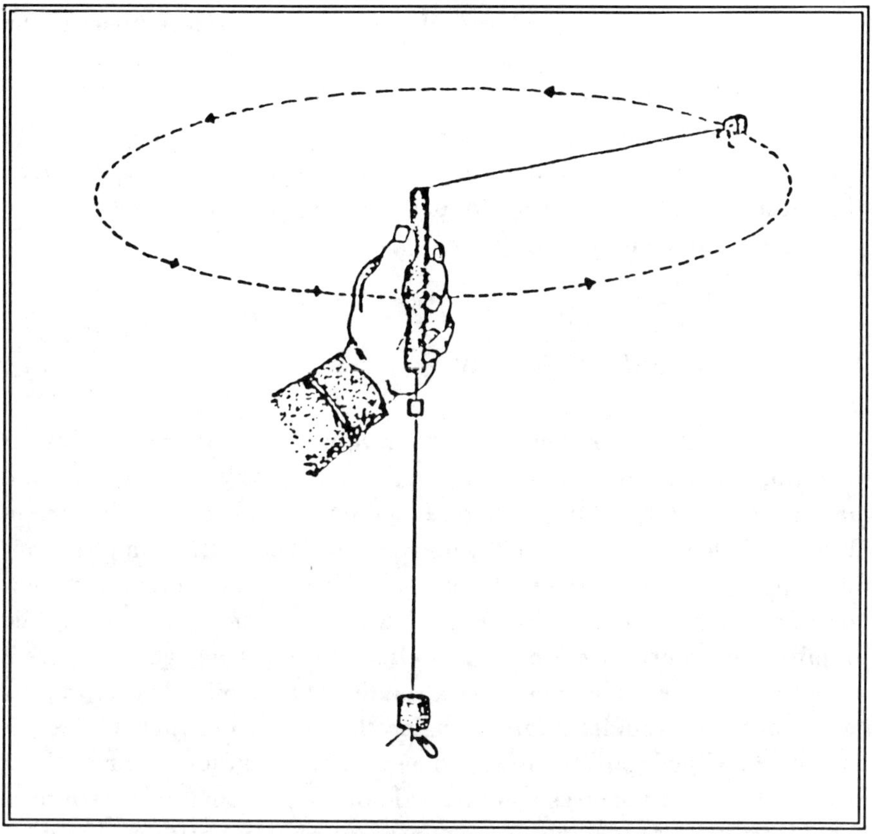

Fig. 4.13. Apparatus for Experiment 4–1.

remains near the bottom end of the glass tubing, thus keeping the radius a constant 100 cm. Remember, the mass of the stopper remains constant.

Determine the time (period) for one revolution (T) to support the number of washers. Repeat T determination by adding various numbers of washers. Plot on a graph the number of washers as a function of T.

How will you determine T? Hint: Count the number of revolutions of the stopper for some set time period, say ten seconds (*frequency*). The period (time) T of one revolution becomes $1/f$. To find T in each case, divide the number 1 by the frequency.

What does your plot of the centripetal force as a function of T look like? Compare the shape of your curve with others in the class.

Now try plotting *F* instead of *T*. How does the *f* curve compare to the *T* curve? What can you say about the relationship of centripetal force and *F* (which is the #/*T*)?

Try using the same number of washers to check the frequency with three or four different radii. Can you determine a relationship between the radius of the path of the stopper and the frequency? Between the centripetal force and the frequency?

*****

Science/Technology/Society

What are some implications of Newton's laws in regard to your transportation needs? A very large automobile may be luxurious and have a wonderfully comfortable ride, and owning one may have tremendous ego-elevation aspects for you. But purchasing it has implications in respect to its impact on society and the environment. Remember, acceleration is directly proportional to mass. A massive automobile requires considerably more force applied to the driving wheels to get it to accelerate the same amount as a smaller automobile. Therefore, the larger auto will require a larger engine that uses more fuel. The result is that you will be required to spend more money on fuel. You may think it is worth it, but it causes a heavy drain on available natural resources. As a result of the higher fuel consumption of the larger, heavier automobile, more pollutants are placed in the atmosphere. Also, a greater number of resources are required in the manufacture of the larger auto.

Soil erosion presently is a major concern in the United States. Tons of precious soil are annually swept to the seas. Soil particles of large sizes rather than small sizes are less easily moved by wind and water because of their inertia and allow for greater water percolation into the subsoil. Farming techniques such as reduced tillage, no-tillage or minimum tillage capitalize on appropriate soil size to reduce erosion.

Pollutants can be separated from water through the use of a centrifuge that operates on the principle of centripetal force. The pollutants are pulled into a circular motion, while the water is allowed to continue in straight-line motion to return to lakes and streams.

Several chemical tests are dependent upon the use of a centrifuge. Such tests are crucial in determination of pollutants present in drinking water and water which is to be returned to lakes and streams.

The mass spectrometer can be used to determine the presence of minute quantities of materials in extremely small samples on the basis of their respective masses. Such devices can identify materials that can evade other, less sensitive tests.

Automatic washing machines use a spin-dry cycle to remove water from clothing. The centripetal force of the washing machine basket pulls the clothing into a circular path while the water, due to its inertia, continues in a straight line into the drain pump.

Problems

1. List four examples of inertia that you experience in everyday life.
2. What is the acceleration of a 10-kg body when a force of 10 newtons is applied to it?
3. How much force is required to accelerate a 50-kg mass 10 m/sec^2?
4. A force applied to a 75-kg mass accelerates it 2 m/sec^2. If the same force is applied to a mass of 45 kg, what will be the acceleration on the 45-kg mass?
5. What is the final velocity of a body which has an initial velocity of 10 m/sec and is accelerated 5 m/sec^2 for a period of 10 seconds?
6. Compute the momentum of a body that has a force of 20 nt applied to it for a period of 20 seconds of time.
7. A force of 15 nt is applied to a mass for 30 seconds of time, resulting in a final velocity of 2 m/sec. What is the mass of the body?
8. A 20-kg mass is swung in a circle at the rate of 25 m/sec on the end of a piece of string 1 meter long. What is its centripetal force?

9. Give at least two examples which demonstrate that centripetal force is along the radius of rotation.
10. In Figure 4.13, in what direction is the centripetal acceleration taking place?

Chapter 5

Work and Energy

Work

Work may be described in a variety of ways. For example, some consider holding a heavy mass at extended arm's length to be work. Others describe thinking or mental problem-solving as work. In the physical sciences, however, work involves a force moving an object through a distance. If the object does not move, no work is done on it. When you hold an object at extended arm's length there is "internal" work – e.g., your heart pumps blood through the arm and food is metabolized. But if the object does not move as you hold it, no work is being done on it.

The force considered here is that which moves parallel to the direction the object moves. For example, if you are pushing on a lawn mower to make it move (Fig. 5.1), you are pushing down (P_v) as well as forward (F_h) when you apply the total force (F_t).

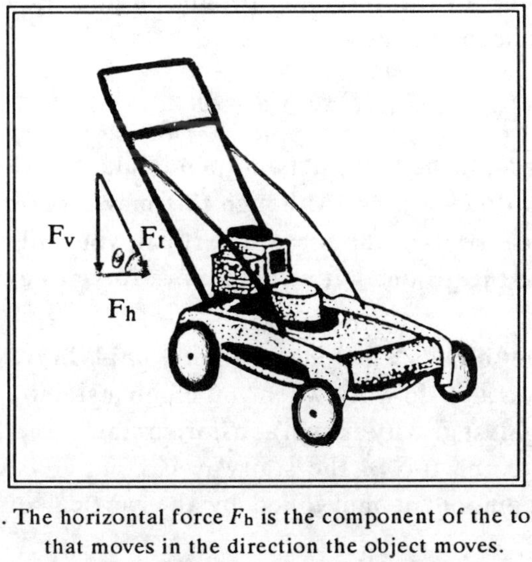

Fig. 5.1. The horizontal force F_h is the component of the total force that moves in the direction the object moves.

The work done on the lawn mower is $F_h \times d$, because the horizontal force, F_h, is the only component that causes forward motion. Usually, you can measure only the total force (F_t) and the angle which F makes with the direction of travel. F_h can be determined easily using trigonometry and is represented by the relationship: $F_h = F_t \times$ *cosine of angle* θ. So the relationship for work becomes:

$$W = F \times \cos\theta \times d$$

The cosine of angle θ, which always ranges from values of zero to one, gets larger as angle θ gets smaller. When angle θ becomes zero, all of the force is horizontal and, as in the example of the lawn mower, the cosine of angle θ equals 1.0. Thus, the relationship becomes:

$$W = F \times 1.0 \times d = Fd$$

When the total applied force is in the direction of the movement of the body, the relationship of the work to the force and the distance becomes:

$$W = F \times d \text{ or } W = Fd$$

Conversely, if the angle θ becomes 90°, all of the force is vertical. It acts perpendicular to the direction of motion and the $\cos\theta$ becomes zero, with work becoming zero:

$$F \times 0 \times d = 0.$$

In other words, if the force acts perpendicularly to the direction of travel, the work done is zero. Although this may seem strange, it does make sense, because if all the force is vertical, you will be pushing the lawn mower into the ground – it will not move forward and no work will be done.

Of course, not all work is done in a horizontal direction as with the lawn mower. You also do work when you climb a stairway by exerting a vertical force against gravity as well as a horizontal force. The work done in raising you to the top of the stairway is computed using only the vertical force component multiplied by the vertical distance moved.

Work is always computed by multiplying the component of force parallel to the direction of motion by the displacement along that component.

In the English system of measurement, the force units are *pounds* and the basic unit of distance is the *foot*. Work is determined in units called *foot-pounds* (ft-lbs). A force of one pound acting through a distance of one foot yields one-foot-pound of work. Only metric units are used in science, but the foot-pound is introduced here to help you understand the historical development of power units treated later in this chapter.

Two units of work are designated in the metric system, depending on the magnitude of the forces and their displacements. The unit of work in the CGS system (centimeter-gram-second) is the *erg* and in the MKS system (meter-kilogram-second) is the *joule* (pronounced "jewel").

An erg of work is done by a force of one dyne moving an object through a distance of one centimeter. Thus, an erg is equal to a *dyne-centimeter*. The *dyne* is the force required to accelerate one gram of mass 980 centimeters per second for one second of time.

A force of one *newton* acting through a distance of one meter results in one joule of work. Consequently, a joule of work is a *newton-meter* of work, where the newton is the force required to accelerate one kilogram of mass one meter per second of time. For example, if a force of ten newtons acts through a distance of one meter, ten joules of work are done (Fig. 5.2). The MKS system is used more frequently because the forces dealt with in nature more closely fit it than the CGS system.

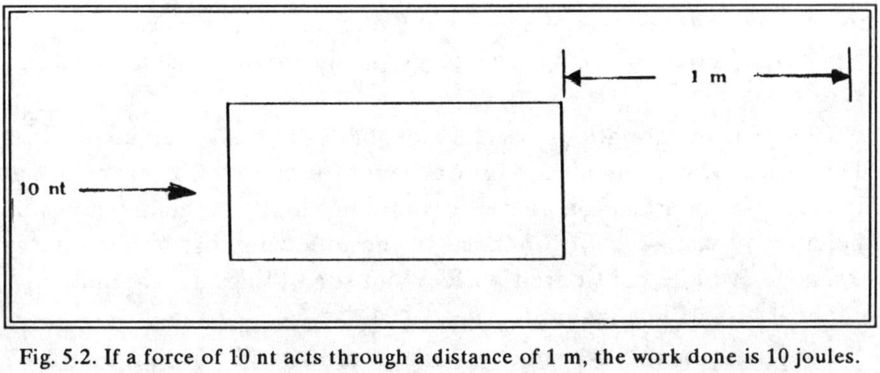

Fig. 5.2. If a force of 10 nt acts through a distance of 1 m, the work done is 10 joules.

Energy

Energy is the amount of work a system has the potential to perform due to its motion, position or condition. The amount of energy a body possesses is due to its position in a gravitational field, velocity, electrical charge or chemical composition. Certain substances made of atoms with naturally unstable *nuclei* (centers) spontaneously give off "nuclear" energy.

Energy which is "stored" or has the ability to do work is *potential energy* (P.E.). The potential energy a body possesses is measured in respect to some point of reference or condition. For example, the gravitational potential energy of a body near the earth's surface is in reference to its distance away from the earth's surface and can be determined by computing the work required to move it to its position. If you climb a set of stairs (Fig. 5.3) to a height of four meters, you have done

$$W = 50\,kg \times 9.8\,nt/kg \times 4\,m = 1960\,joules$$

of work, assuming you have a mass of 50 *kg.* You have gained 1960 *joules* of potential energy due to your position above the earth. The factor of 9.8-nt/kg is needed to convert *kilograms* (units of mass) to *newtons* (units of force). If you slipped and fell to the ground, you would accelerate due to gravity until you reached a maximum velocity as you hit the floor. Because you have a velocity, you now possess the energy of motion which is labled *kinetic energy* (E_k). At the instant you strike the floor, all 1960 joules of potential energy exist in the form of kinetic energy (E_k) and potential energy is zero. The kinetic energy is transferred to the floor upon impact in the form of heat – not a pleasant way to warm a floor. It is perhaps better to use spheres or bodies in the discussion of energy transformation.

Energy transformations may be thought of as a bookkeeping process. Theoretically, you can always fully account for energy. The transformations may be in areas you normally would not look for them, but nevertheless, you can account for them. In the following discussion, energy exchanges will be considered perfect and such things as frictional loss and heat loss will be ignored.

Suppose you roll a perfectly smooth and round bowling ball up a perfectly smooth hill and allow it to roll down the frictionless incline as

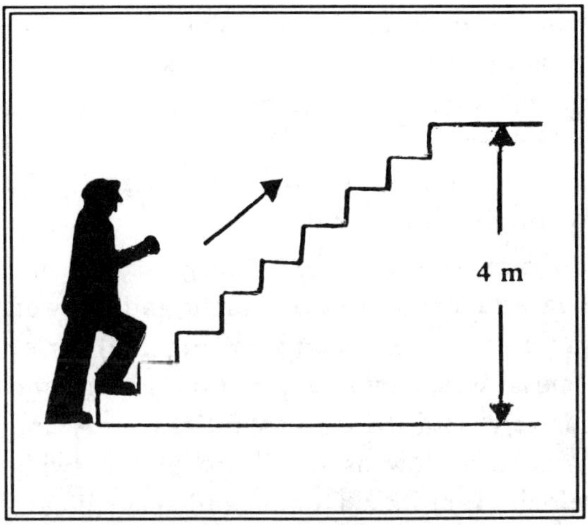

Fig. 5.3. A 50-kg person gains 1960 joules of potential energy climbing a 4-m stairway.

illustrated in Figure 5.4. Because you exert 10 nt of force through a vertical distance of 10 m, you do work on the ball of:

$$10\,nt \times 10\,m = 100\,joules$$

Thus, at the top of the hill the ball possesses 100 joules of potential energy. If you allow the ball to roll down the incline from the top of the hill, it loses potential energy as it gains kinetic energy. At the bottom of the incline all of the potential energy has been transferred to kinetic energy. The ball's potential energy is now zero and its kinetic energy is a

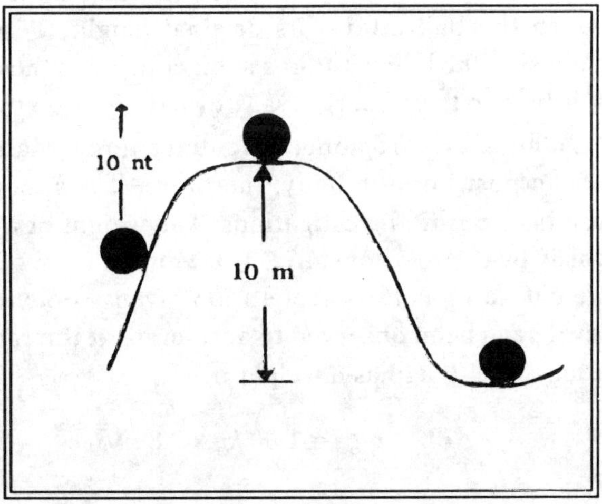

Fig. 5.4. It takes 10 nt of force to roll a bowling ball up a hill to a height of 10 m.

maximum 100 joules. As the ball continues up the next slope it begins to slow down and lose its kinetic energy. At the same time, it is doing work against the force of gravity and gains potential energy. Because its mass remains constant, gravity will affect the ball the same amount you affected it by rolling it to the top of the hill. Thus, it rises to a height exactly the same as the heigth to which you originally rolled it.

Perhaps you have seen a demonstration where a large, heavy ball attached to a long rope is released from a specific height and allowed to swing toward the head of a person across the room (Fig. 5.5). Prior to its being released, the ball possesses maximum potential energy and zero kinetic energy. After it is released, it loses potential energy and gains kinetic energy, until it goes as low as it will go at the swing midpoint. At that point its velocity is at its maximum and according to the relationship

$$Ek = \frac{mv^2}{2} = \text{maximum}$$

After passing the midpoint, the ball continues toward the subject and begins to lose kinetic energy as it simultaneously gains potential energy. Gravity continues to decelerate the ball until its kinetic energy becomes zero and thus its velocity is zero. At that instant the ball is at rest, and, if the subject is in the appropriate place, he/she is spared being struck by the ball. The trick is to measure the height from which the ball is released and be certain the cooperating subject is beyond the point at which the ball attains its original height. Test your measurements by releasing the ball without a subject prior to the demonstration. (Perhaps it is best to trust energy transformation or do the test without a subject.)

One final word about potential energy. Mass is consistent regardless of the position of a body; therefore, it is a more dependable entity for use in scientific investigations. The weight of a body can be determined easily by the relationship $F = ma$ in the form of $W = mg$, where W is the weight and g is the acceleration given a body by gravity. Bodies at sea level have been observed to accelerate at the rate of 9.8 m/sec/sec. Thus, a mass of 1.0 kg has a weight of:

$$W = mg = 1.00\ kg \times 9.8\ m/sec^2 = 9.8\ kg\ m/sec^2$$

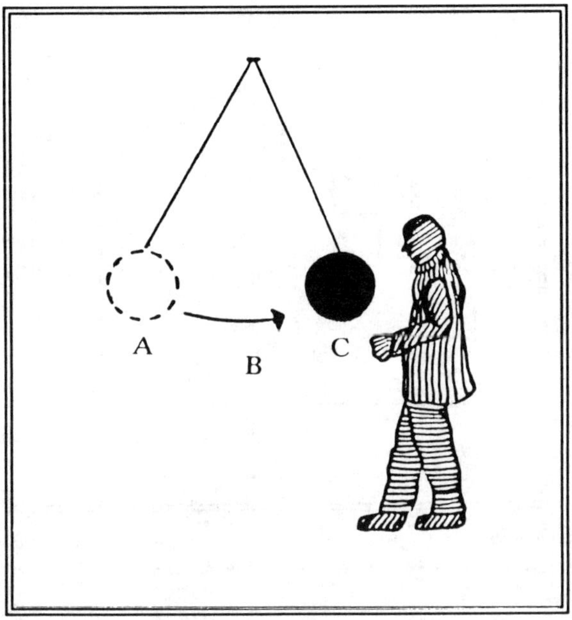

Fig. 5.5. The ball loses all of its potential energy at point "B" and regains it all again at "C."

Because 1 *kg* m/sec^2 has been defined as a *nt*, 9.8 kg m/sec^2, a mass of 1.0 kg at sea level has a mass of 9.8 nt.

Power

Power is a rate of doing work. Doing a certain amount of work in a specific amount of time is an example of power. James Watt originally attempted to develop a unit of power by basing it on a common source of energy, the horse. He estimated that a horse could do 550 ft-lbs of work in one second of time and named the unit one horsepower. Horsepower is presently determined in a variety of ways for engines of all sorts.

The unit of power of interest in the metric system is the *watt*, named in honor of James Watt. A joule of work done in one second of time is a *watt of power*. Thus a watt is a joule per second. To compute the power, first compute the work done and divide it by the time in seconds.

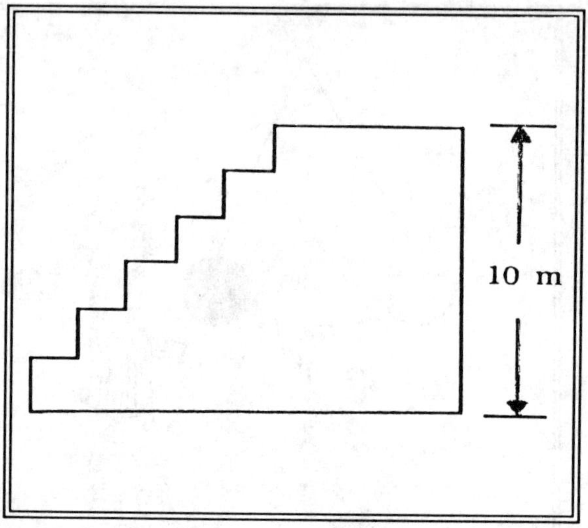

Fig. 5.6. To compute one's power, the vertical height needs to be measured. The height of one riser can be multiplied by the number of risers.

You will find it interesting to compute the power a person can generate, and stairsteps are a convenient place to make this determination. The time required to run to the top of the stairs or series of stairs, the vertical height of the stairs in meters and the weight of the person in newtons are what are needed to compute the power.

The following calculations, using the height of the stairs in Figure 5.6 and a 70-kg man who ran up the stairs in 22 seconds, serve as an example:

mass of man	= 70 kg
height of stairs	= 10 m
time	= 22 sec

$$Power = 70\ kg \times \frac{9.8\ nt/kg}{22\ sec} \times 10\ m = \frac{311.8\ nt\ m}{sec}$$

$$= \frac{311.8\ joules}{\text{sec}} = 311.8\ watts$$

A man of 70-kg mass running to the top of a stairway of vertical height of 10 m in 22 seconds generated 311.8 watts of power!

If you are interested in the horsepower generated by the man, divide the watts by 746 *watts/hp*:

$$HO = \frac{311.8\ watts}{746\ watts/hp} = 0.42\ HP$$

The man is rated at almost one-half of a horsepower. If the power output of the man could be hooked to an electric generator, he could light a 311.8 watt bulb. The question is how long he could maintain this power output. Most likely the bulb would quickly dim and soon be extinguished.

It can be seen that the unit *watt-hr* of electricity is not a unit of power but rather a unit of energy because

$$1\ watt\text{-}hr = \frac{1\ joule\text{-}hr}{sec}$$

and time appears both in the numerator and denominator. Thus, time divides out of the equation, leaving joules, which are units of work or energy. An electrical power company is in reality an energy company.

Science/Technology/Society

A catchphrase in vogue today is the term *renewable energy resource.* The term really refers to short-term, renewable energy sources. Coal and petroleum, for example, are fossil fuels derived from organisms from a previous era that depended on the sun for energy. Theoretically, given enough time and appropriate conditions, coal and petroleum are renewable resources. But the reserves in use today were formed millions of years ago, and an industrial society needs energy sources that can be produced in a matter of days.

One such renewable energy source is ethyl alcohol (*ethanol*), which is derived from the fermentation of grains, tubers and fruits that store the energy of the sun. Actually, solar energy is converted to potential energy. If it is chemically unlocked, this potential energy can be converted to kinetic energy to, for example, turn the wheels of an automobile. Ethanol has other benefits, in that its combustion produces carbon dioxide and water but does not produce nitrogen and sulfur compounds as does the combustion of petroleum and coal. The most successful application of ethanol as an energy source has been through mixing it with gasoline to form a product called gasahol.

Unfortunately, ethanol from grains has not been able to compete commercially with gasoline. When gasoline was well above $1.00 per gallon, there was considerable interest in burning ethanol as a petroleum substitute for energy. The value systems of people are such that there continues to be a dependency upon petroleum as a major energy source. Even some grain farmers react against using alcohol as fuel in their vehicles. Find out reasons why farmers do not universally accept and promote ethanol as a viable energy source.

Wind is another form of available energy, though it is not particularly dependable. Certain areas have more consistent wind patterns than others. One example is the north shore area on the Hawaiian island of Ohau. Here, electricity is being produced on a commercial basis in a "wind farm." When considering wind as a renewable energy resource, you need to consider the following questions: What causes wind? Can wind be considered a form of solar energy? In what respect?

Trees are also renewable energy sources. However, trees take a relatively long period of time to mature and require large land masses. Trees also have considerable aesthetic value and people are sensitive about their destruction. Trees, such as Christmas trees and those grown for lumbering purposes, are presently grown on a commercial basis.

One clean, convenient source of renewable solar energy is hydroelectric energy. Hydroelectric energy is a form of solar energy. The water is given its potential energy by the sun. Water evaporated by the sun rises into the atmosphere and descends to the mountain tops in the form of snow which is melted by the summer sun. For example, a series of dams on the Missouri River consistently produces a large block of electrical energy. Runoff snow-melt from the Rocky Mountains and the rain and snow from the Northern Plains flow into the Missouri River as it proceeds toward the Gulf of Mexico. The dams raise the water level, increasing the potential energy of the water in that specific location and that energy is converted to kinetic energy. The process continues as long as the sun provides the energy.

Contrary to the beliefs of too many people, present energy sources are not unlimited. As the population increases geometrically and yearly per capita energy demands increase, available energy supplies are shrinking at an alarming rate.

It is becoming increasingly important that individuals conserve energy. Waste is the enemy of energy sources. The policies of a throw-away society and planned obsolescence result in unnecessary energy waste. Present-day value systems will need to change so that people make decisions more on the basis of minimal energy consumption than they have in the past.

It is important to remember as the consumption of "dirty" fuels, such as coal and petroleum, increases, the amount of resultant pollution also increases – a two-edged sword. One problem generates or increases one or more other problems.

Problems

1. How much work does a boy do in holding a 25-kg mass at a height of 2 m above the floor for a period of 25 minutes?
2. A 441-nt girl walks up 50 steps of a staircase. If each riser of the staircase is 14 cm high, how much work did the girl do?
3. A weightlifter presses a 1764-nt barbell to a height of two meters 10 consecutive times. How much work did she do in raising and lowering the barbell?
4. Suppose you exert 500 nt of force in pushing an automobile a distance of 100 meters. What is the amount of energy (work) you expend?
5. If you exert 20 nt of force in sliding a 500-nt box 10 m to the back of a truck and then lift the box up 1 m to the bed of the truck, what is the total work expended?
6. What is the potential energy given to a 24-kg concrete block in lifting it to a position in a wall at a height of 3 meters? (Assume each kilogram of mass exerted has a gravitational force of 9.8 nt acting on it.)

7. If a 20-nt ball is rolled from the top of a hill of vertical height of 5 meters, and comes to rest at a height of 5 meters on the opposite hill that is 15 m high. (a) What is its potential energy at rest? (b) How much potential energy did it lose? (c) How much work would be done in moving it to the top of the hill?

8. A 15-kg ball on the end of a rope is pulled aside so it is 0.25 meters above its central resting point. (a) What is its gain in potential energy (1 kg = 9.8 nt)? (b) What is its kinetic energy midway through its swing? (c) What is its velocity midway through its swing? (d) What is its potential energy at the far end of its swing? (e) What is its kinetic energy at the far end of its swing?

9. A student weighs 80 pounds. If the student runs to the top of a 10-m flight of stairs in 5 seconds of time, how much power was developed (1 kg = 2.2 pounds)?

10. If the energy of a 220-lb boy who runs to the top of a 10-m stairway in 5 seconds could be converted directly to electrical energy, what size light bulb could be lighted with the energy generated?

Chapter 6

Heat and Some Heat Applications

The Nature of Heat

Thermal Expansion of Solids

In general, substances become larger (expand) when heated and smaller (contract) when cooled. Gases are affected more by heat than solids or liquids. In this section you will explore whether the thermal expansion of a solid is a characteristic property.

You would expect that, if a rod of material expands a specific amount in length when heated, an identical rod of the same material would expand the same amount if heated to the same temperature. Likewise, if two identical rods are placed end to end and heated, the increase in the length would be twice that of the increase in the length of a single rod. If two rods are placed side-by-side, the change in length would be the same and equal to the change for one rod. You would not expect thickness to affect the length (see Fig. 6.1).

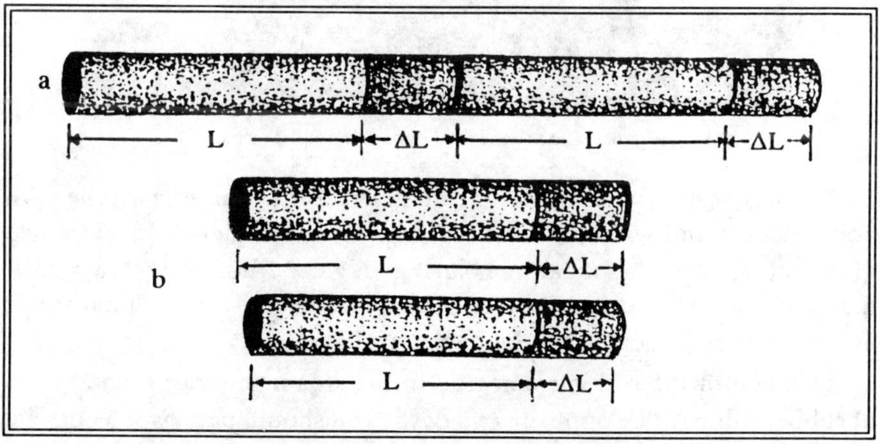

Fig. 6.1. (a) Two identical rods placed end to end expand twice as much (2 ΔL) as a single rod (ΔL) when heated; (b) two rods placed side by side expand only 1 ΔL when heated.

*****

Experiment 6–1. Thermal Expansion of Solids

In this experiment you will use the apparatus shown in Figure 6.2 to measure how much the length of a rod changes when heated to the boiling temperature of water (the linear thermal expansion). You will be investigating several hollow rods of the same diameter and length but of different materials, i.e., aluminum, copper and glass. A second aluminum rod, identical to the other rods except for a larger diameter, will be used to determine whether diameter affects thermal expansion, at least for aluminum.

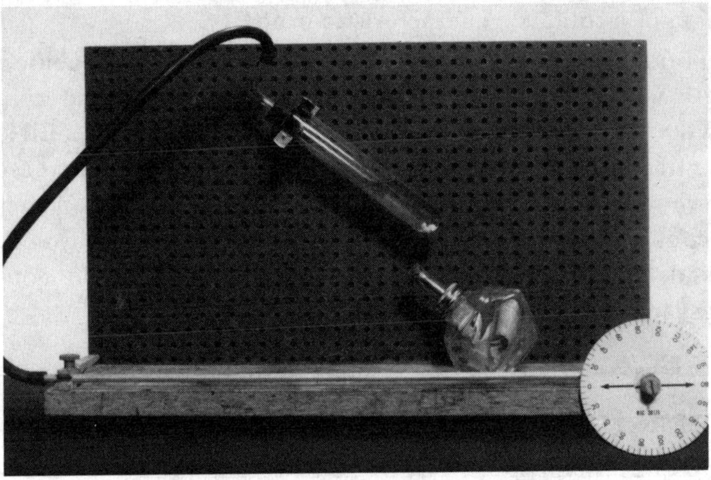

Fig. 6.2. Linear thermal expansion apparatus.

The apparatus consists of a steam generator (a large test tube with boiling chips and water, a one-hole stopper and a piece of right angle glass tubing) and a linear measuring device. First, select a small-diameter, hollow rod and firmly fasten it at one end with a thumbscrew and a clothespin located on the pegboard base.

This end of the rod is connected to the steam generator via a piece of rubber tubing. The opposite end of the rod should pass over a corsage pin laying on a piece of emery paper glued to a masonite block. A rubber band firmly holds the corsage pin against the emery paper. The opposite end of the corsage pin is fastened to a cardboard dial (Fig. 6.3).

Fig. 6.3. Dial apparatus.

The dial is calibrated in angular divisions. As the rod becomes heated, it expands away from the fixed clothespin and moves over the corsage pin, causing the pin to rotate. The cardboard attached to the pin also rotates, but due to its much larger diameter, amplifies the movement of the pin so a very small movement can be measured readily. Thus, you can record relatively small amounts of linear expansion of the rods.

Prior to applying heat to the steam generator, the dial should be "zeroed in" with the edge of the base of the supporting board, which serves as a reference point. As steam flows through it, the rod becomes increasingly heated and eventually acquires the same temperature as the steam. At this temperature, linear expansion stops and the dial maintains a maximum reading. Record the maximum reading.

Repeat the process with each of the rods. It is important that the apparatus is set up exactly the same for each rod. Make sure the distance from the clothespin fastener to the corsage pin is the same for each rod, as that determines the length under investigation. What happens if this length varies from rod to rod? Record and compare the readings on the dial for each rod. How do the readings for the aluminum rods of differing diameters compare? Does the diameter make a difference?

You may graph the angular readings from the dial in the form of a histogram for each rod. Is linear thermal expansion a characteristic property of the material?

*****

The Coefficient of Linear Expansion

In the preceding experiment you compared the linear expansion of materials from room temperature to the boiling temperature of water. You could have placed the rods in a container of water and heated them over a whole range of temperatures to determine the amount of change in length at each temperature.

For applications that involve large pieces of material, such as a bridge span, the above method would not be used to find how much it will expand. But if you know how much each unit will expand for each degree temperature change, the total expansion of the piece can be calculated. This is called the *coefficient of linear expansion* because it is a ratio or fraction of the original length. For example, if you heat an aluminum rod 50 cm long from 25°C (about room temperature) to 30°C, it increases in length 6.25 x 10 cm^{-3}(Fig. 6.4). You can compute the ratio of the change in length $\Delta L/\Delta L$:

$$\frac{\Delta L}{\Delta L} = \frac{6.25 \times 10\text{-}3\ cm}{50\ cm} = 1.25 \times 10^{-4}$$

Notice that the units divide out, thereby providing a unitless ratio that can be applied to any linear measurement.

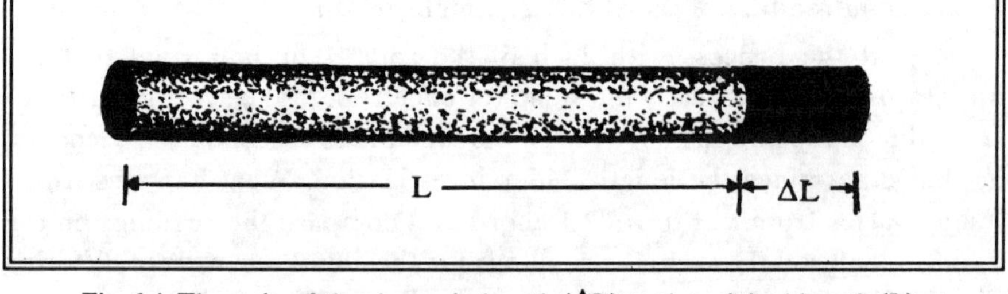

Fig. 6.4. The ratio of the change in length (ΔL) to the original length (L).

The temperature change is considered by dividing the ratio of change in length by the temperature change. In this example, the temperature was raised from 15°C to 30°C, resulting in a temperature change of 5°C. Dividing the ratio of "change in length" (0.92×10^{-4}) by the temperature change of 5°C yields:

$$\frac{1.25 \times 10^{-4}}{5°C} = \frac{0.25 \times 10^{-4}}{°C} = \frac{2.5 \times 10^{-5}}{°C}$$

The *coefficient of linear expansion* for aluminum is $1.8 \times 10^{-5}/$ °C. Each degree Celsius temperature change will alter the length of an aluminum piece 1.8×10^{-5} units for each original unit of length. For example, an aluminum rod 10 m long will change in length when heated from 25°C to 40°C:

$$\Delta L = \frac{10 \text{ m} \times 2.50 \times 10\text{-}5 \times (40\,°C - 25\,°C)}{°C}$$

$$= \frac{10\, m \times 2.50 \times 10^{-5} \times 15\,°C}{°C}$$

$$= 3.75 \times 10^{-3} m = 3.75\, mm$$

The rod will become 3.75 mm longer due to heating. The coefficient of linear expansion is slightly dependent upon the temperature range through which it is determined, but for most uses, this anomaly is not significant.

Table 6.1 is a list of some common materials and their coefficients of linear expansion. A table such as this could be used in the construction of a building, an overhead walkway or bridge. The periodic bumps you experience on a concrete highway are due to the asphalt being squeezed out of "expansion cuts" placed in the highway to compensate for thermal expansion.

Table 6.1. Coefficients of thermal linear expansion.

Material	$1/°C \times 10^{-6}$	Material	$1/°C \times 10^{-6}$
aluminum	25.0	glass	8.6
brass	19.0	iron	12.0
copper	17.0	pyrex glass	3.1
concrete	12.0	silver	10.0

Thermal Expansion of Liquids

Is the thermal expansion of liquids a characteristic property? Can you distinguish between liquids by studying their thermal expansion? You can attempt to answer these questions on the basis of examining some liquids in the laboratory using the set up in Figure 6.5.

Fig. 6.5. Thermal expansion of liquids apparatus.

The levels of the liquids in the glass tubing must be manipulated to be equal at the beginning of the experiment. The water in the large beaker is heated and the levels are recorded for every 10°C rise in the temperature. Such a set of data are shown in Table 6. 2.

A graph of the data in Table 6.2 can be found in Figure 6.6. An inspection of the graph indicates three separate, distinct plots. On the basis of the changes in liquid levels recorded for these three liquids, there is good evidence that thermal volume expansion is indeed a characteristic property of liquids. The coefficients of volume expansion can also be computed as a ratio of change in volume to the original volume divided by the temperature change. Keep in mind that the measurements described with the apparatus depicted in Figure 6.5 are amplified due to the small diameter of the glass tubing compared to the diameter of the test tube. A

Table 6.2. Height of liquid as a function of temperature.

°C	Height of Liquid (cm) Vegetable Oil	Water	Isopropyl Alcohol
20	0.0	0.0	0.0
25	0.6	0.2	0.8
30	1.2	0.4	2.1
35	2.4	0.9	3.8
40	3.6	1.6	6.2
45	4.8	2.4	8.3
50	6.6	3.4	11.3
55	8.0	4.4	14.0
60	9.4	5.8	17.6
65	11.2	7.4	20.7
70	13.2	9.2	25.2
75	15.1	11.8	37.0

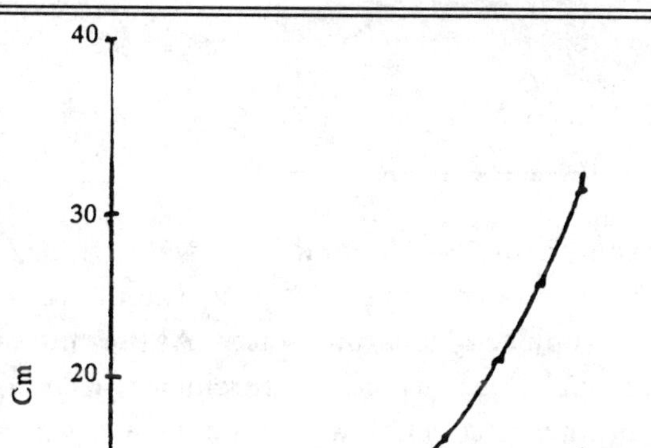

Fig. 6. 6. Graphic representation of data in Table 6.2.

small volume change in the test tube manifests itself as a much larger change in liquid level in the glass tubing.

Water is a liquid which has a non-linear coefficient of volume expansion. The density of liquid water increases as the temperature decreases to 4°C, at which temperature it is most dense. As the temperature decreases below 4°, the density again decreases (Fig. 6.7).

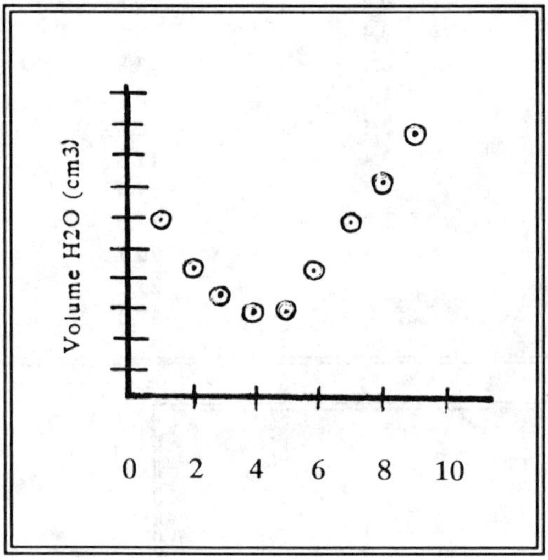

Fig. 6.7. Graph illustrating the expansion of a liquid with a rise in temperature.

This phenomenon explains why bodies of water freeze from the top down. As the surface is cooled toward 4°C, the water at the surface settles to the bottom and is replaced by warmer water. As the entire body of water approaches 4°C through this convection interchange, its density stabilizes. As the surface continues to cool below 4°C, the surface water becomes less dense and, at 0°C, the surface begins to freeze. Thus, biological forms such as fish can survive in the water below the frozen surface.

Thermal Expansion of Gases

You have observed that the thermal expansion of liquids and solids appears to be a characteristic property. Is thermal expansion also a characteristic property of a gas? Recall that if the temperature is kept constant and the pressure is varied, the corresponding volume change is the same for all gases.

You can use the same apparatus to check the thermal expansion as you used for varying the pressure of a gas (Fig. 6.8). The same mass is kept on the plunger while the temperature of the gas is varied by immersing the syringe in a water bath.

After you have recorded the data corresponding to periodic temperature changes for one gas, you can flush out that gas, replace it with another gas and repeat the process. A plot of the data shows the curves to be coincident. Analysis of the data plots indicates that gases behave pretty much the same as the temperature is varied and the pressure maintained. Thermal expansion of gases does not seem to be a characteristic property. You would not depend on thermal expansion to help you identify a gas.

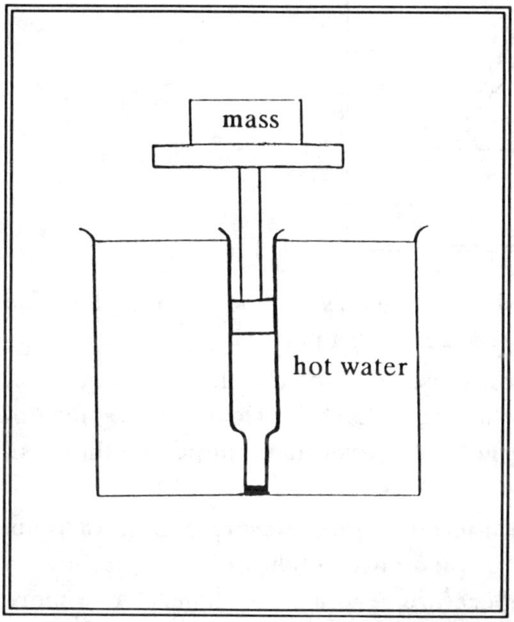

Fig. 6.8. Apparatus used to check thermal expansion of a gas.

*****

Experiment 6–2. Boiling Temperature

Boiling water is a fairly common phenomenon. It is well known that water boils at a specific temperature (100°C). Other materials can also be made to boil. Even a hard substance such as steel can be made to boil

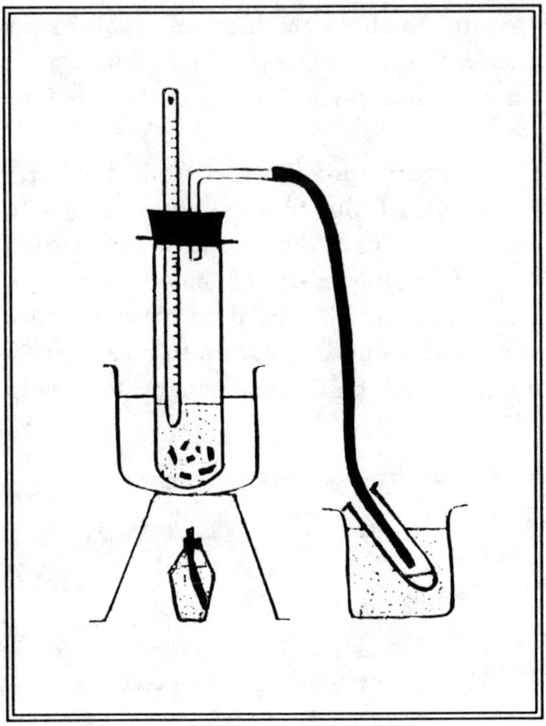

Fig. 6.9. Boiling temperature apparatus.

if heated sufficiently. The question is: Do all substances boil at the same temperature or is boiling a characteristic property?

Place a few small boiling chips in a test tube and add about 10 ml of a liquid. Place a thermometer in one hole and a right-angled piece of glass tubing in the remaining hole of a two-hole stopper as shown in Figure 6.9.

Heat the liquid, recording the temperature every minute of time. Note the time and temperature when boiling takes place. Continue to heat the liquid and take temperature readings for at least five more minutes after it begins to boil. Graph the data. What does the graph look like after boiling begins? At what temperature did the liquid begin to boil? Compare your graph with those developed by students who used different liquids? Is the boiling temperature of a liquid a characteristic property?

*****

Experiment 6–3. Freezing/Melting Temperature

Most substances have a melting temperature. A welder uses an oxygen-acetylene torch to melt steel into a puddle. When the torch is removed, the steel solidifies (*freezes*). While freezing is the process of changing a liquid to a solid, the term is usually reserved for temperatures well below room temperature.

In this experiment you will turn a substance first into a liquid and then allow it to cool until it solidifies completely. The experiment does not actually begin until all of the material is melted.

Place about an inch and a half of either moth flakes or moth balls into a test tube. Stand a thermometer, with a slotted cork stopper slipped over it, on top of the material as in Figure 6.10. The test tube and a second thermometer (optional) are supported in a beaker of water that will serve as the hot water bath to yield a more gradual temperature change.

Heat the water while occasionally stirring the material in the test tube, using the thermometer as a stirring rod. Just before all of the material is melted, remove the burner. The residual heat of the water will complete the melting.

Fig. 6.10. Apparatus for exploring the freezing/melting temperatures of a substance.

With the burner removed, record the temperature for both the water and the liquid in the test tube at one minute intervals (Record the starting time as 0). Note the time solidification begins and is completed.

Graph the temperature (vertical axis) as a function of time (horizontal axis). Note any patterns or perturbations in the curve. At what temperature did "freezing" begin? At what temperature did it stop? Can you identify this phenomenon on the curve plotted on the graph?

Compare your graph with those of others who worked with different materials. How is your graph the same? How is it different? Would the curve for water look different? To what temperature would you need to extrapolate the graph to show the freezing temperature of water? Why wasn't water used in this experiment instead of the material you used? The process is reversible, as can be demonstrated by heating the material you used until it melts, checking and recording the temperature each minute. In other words, you could predict from the graph how the substance would behave as it warms, begins to melt, completes melting and continues to increase in temperature. All you need to do is follow the curve of the graph in reverse of the process you used in the experiment. Freezing and melting occur at the same temperature!

*****

Measures of Heat

Temperature and heat, although related, are not the same. Actually, temperature is the average kinetic energy of the particles that make up a substance. Temperature can be looked at as a relative measure. It is a comparison to some physical phenomenon such as the freezing temperature of water or absolute absence of energy (*Kelvin temperature scale*).

Heat (*thermal energy*) takes into account the total energy a substance possesses. If two identical pans, the first one-fourth full of water at 30°C and the second pan three-fourths full of water also at 30°C , have the same intensity of heat applied to them (Fig. 6.11), the pan with the smaller amount of water would reach 50°C much more rapidly than the pan with three times the amount of water.

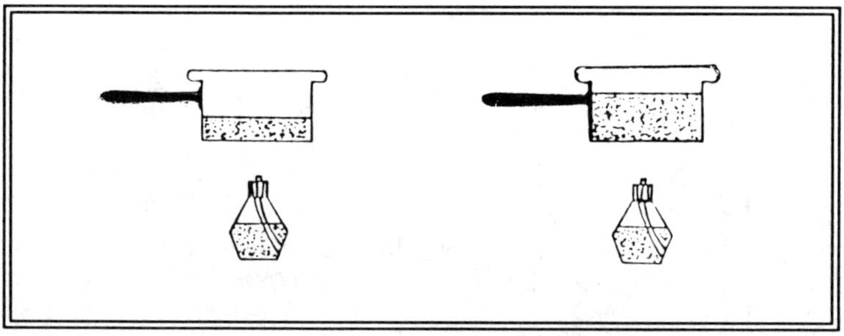

Fig. 6.11. The pan on the right contains three times the amount of water at 30°C.

Two units of measure are the *calorie* and the *kilo-calorie*. By definition, the *calorie* is the amount of heat required to change the temperature of one gram of water one degree Celsius. The *kilo-calorie* is the amount of heat required to change the temperature of one kilogram of water one degree on the Celsius scale: 1 Kg = 1000 *grams*, 1 kilo-calorie = 1000 calories. Sometimes the lower case letter *c* is used as the representation of the calorie and the upper case letter *C* to represent the kilo-calorie. Dietitians most often use the kilo-calorie. Although either unit could be used, the calorie involves smaller numbers.

Suppose a piece of marshmallow is completely burned underneath a test tube containing 50 grams of water at a temperature of 30°C (Fig. 6.12). After the marshmallow is completely burned, the temperature of the water is observed to be 90 °C. If you assume that all the heat from the marshmallow is transferred to the water and the test tube itself absorbed no heat, you can compute the amount of heat transferred to the water. This is the amount of heat given off from the combustion of the marshmallow:

$$\text{Heat} = \frac{50\,g \times 1\,Kg \times (90\,°\text{C} - 30\,°\text{C})}{1000\,g}$$

$$\frac{1 \times 60\,Kg\,°\text{C}}{20} = 30Kg\,°\text{C} = 30\,K\text{-}cal$$

If you had eaten the piece of marshmallow, you would have made 30 kilo-calories of energy available to your body. Should your metabolism not require 30 more kilo-calorie of energy to maintain the status quo, it most likely would squirrel it away as fat.

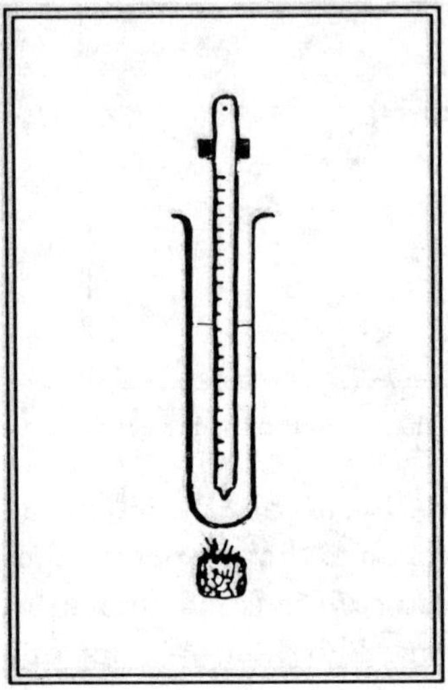

Fig. 6.12. Determining calories by burning a marshmallow.

Specific Heat

The preceding discussion described the transfer of the heat from a burning marshmallow to a mass of water. Remember, the assumption was that all of the heat was transferred to the water. Even allowing for zero heat loss to the atmosphere, all is not so simple in nature. The water must be held in a container – the test tube – so it, as well as the thermometer, will absorb some heat.

In order to account for all of the heat transfer, one must determine the mass of the glass and the heat capacity of each kilogram of glass. It turns out that the heat capacity of glass is about 0.2 that of an equal mass of water.

Specific heat is the measure of the amount of heat each unit of mass of a substance can absorb for each °C change in temperature in comparison to a standard substance. Because water was used for the basic definition of the kilo-calorie, it is conveniently used as the standard for specific heat. The specific heats of all substances indicate how their heat capacities compare to that of water. For example, glass has a specific heat of 0.2, so its heat capacity is only 0.2 kilo-calorie per kilogram per degree Celsius or 0.2 times the heat capacity of water.

The impact of the specific heat of a substance can be observed in weather patterns. If you live near a very large body of water such as the

Great Lakes or an ocean, the temperatures do not fluctuate as much as if you lived on the Great Plains. The specific heat of the land mass is not as great as the specific heat of water; therefore the land mass tends to heat up and cool off more rapidly than water mass.

Heat Engines

The modern standard of living would not be possible without the use of heat engines. Heat engines are used in chainsaws, lawn mowers, weed trimmers, tractors, automobiles, trains, airplanes, ships, rockets and a host of other applications. Until a substitute is found for heat engines, humanity will depend heavily on them to convert fuels into energy. Because heat engines influence so much of your life, it is imperative that you have a basic understanding of how they work.

Steam Engines

James Watt is credited with the invention of the basic steam engine. In the fundamental design, a cylindrical piston moved inside a metal cylinder in which hot steam expanded against the piston causing movement. The moving piston was connected to the work desired.

The Watt steam engine was a reciprocating engine because the piston moved back and forth (reciprocated) in the cylinder. It had a cooling cycle that depended on atmospheric pressure to return the piston, but this feature has been discontinued on modern engines.

A study of the reciprocating steam engine will aid the conceptualization of all reciprocating engines, such as the gasoline engine, diesel engine and, to some extent, the turbine heat engine. The application of the reciprocating steam engine is currently limited to museum exhibitions or tourist attractions. You can recognize the reciprocating engine by the colorful huff-puff sound it makes as the pistons move back and forth, exhausting steam.

Figure 6.13 depicts the reciprocating steam engine. Steam is first directed in front of the piston, forcing the piston back, and the expended steam behind the piston is expelled through the exhaust valve.

The piston rod turns the crankshaft in the same manner the force of your foot turns the pedal (crankshaft) of a bicycle. Steam is then

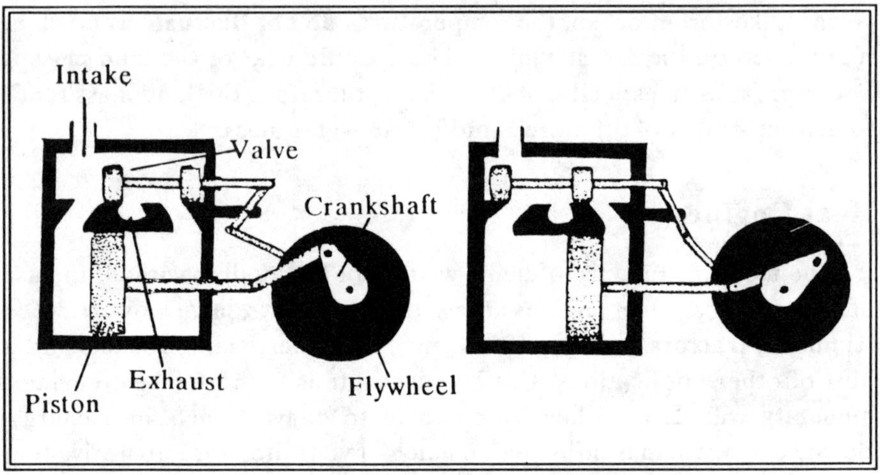

Fig. 6.13. The reciprocating steam engine. Steam is directed in front of the piston as spent steam is exhausted.

directed behind the piston, forcing it forward and expelling the spent steam in front of the piston. A clever valve system provides both an intake path for the steam on one end of the piston and a "spent" steam exhaust path at the other. The valve system is directly connected to an oval-shaped *eccentric* mounted on the crankshaft. The eccentric moves the valve system back and forth at precisely the right time.

The heavy flywheel increases the inertia of the crankshaft to which it is attached. It keeps the crankshaft turning smoothly during periods when the piston stops to reverse its direction of travel. Inertia, you will recall, is the tendency to keep a body moving once in motion. Obviously, a reciprocating engine is not a particularly smooth engine, especially at slow revolutions of the crankshaft. Remember, the piston moves in one direction, abruptly stops, reverses direction, accelerates to maximum speed and then stops and reverses direction again.

Although colorful and romantic, the reciprocating steam engine is not very efficient. A more efficient engine is the steam turbine, such as those used in steam-powered electrical generating plants and sea-going ships. In these engines the steam is directed against turbine blades that are arranged in a circle on the main shaft to which a generator or propeller is attached (Fig. 6.14).

As the steam expands against the blades of the turbine wheels, some of its potential energy is transferred to the kinetic energy of the rotating shaft, similar to the way in which wind energy is transferred to the blades of a pinwheel or an old-fashioned windmill.

Fixed vanes (stators) direct the steam against the turbine wheels at the most efficient angle. The shape of the turbine case, the vanes and the blades are designed for the most efficient transfer of the thermal energy of the steam to the turbine shaft.

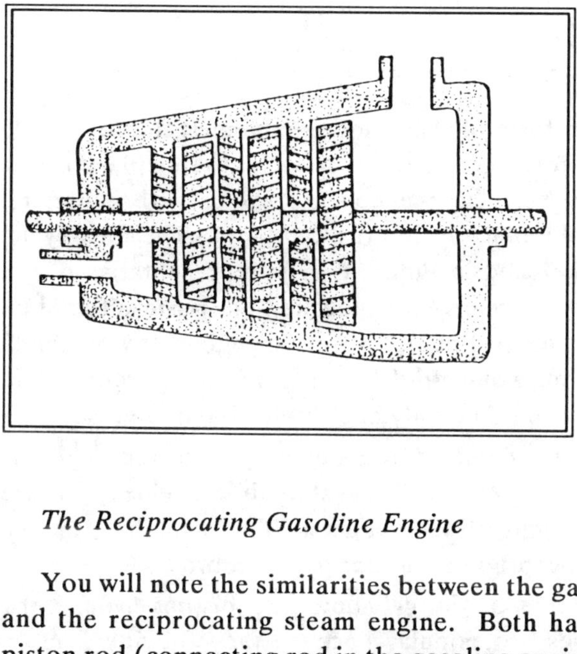

Fig. 6.14. Steam turbine.

The Reciprocating Gasoline Engine

You will note the similarities between the gasoline engine (Fig. 6.15) and the reciprocating steam engine. Both have a piston, crankshaft, piston rod (connecting rod in the gasoline engine) and a flywheel. Both use expanding gases to move a piston in a cylinder.

The gasoline engine differs from the steam engine in several ways. The combustion takes place outside of the cylinder of a steam engine, but inside of the cylinder of a gasoline engine; therefore, steam engines are referred to as *external combustion* engines and gasoline engines as *internal combustion* engines. The internal burning of the gasoline takes place only in front of the piston of a gasoline engine. No energy is applied to the backside of the piston, as in the steam engine.

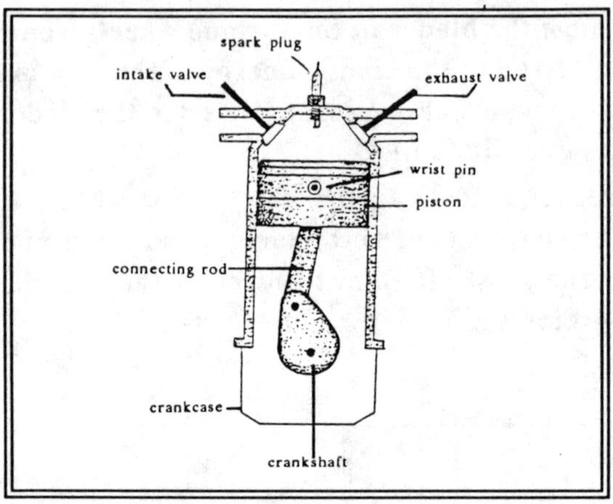

Fig. 6.15. Major components of a gasoline engine.

The most widely used form of the gasoline engine is the four-cycle (four-stroke) engine. Reference is to the four cycles taking place inside each cylinder of an engine. Do not confuse cycles with cylinders. Remember, the cycles take place inside the cylinders. The four-cycle, single-cylinder engine sketched in Figure 6.15 is representative of the single-cylinder engine used in many lawn mowers. The basic automobile engine in use today is the four-cylinder, four-cycle engine. If you looked inside such an engine, you would find four of everything depicted in Figure 6.15, except there would be only one flywheel and crankshaft.

Usually the cylinders are lined up in a single row and the engine is appropriately labeled an *in-line* engine. Automobile engines also are designed with six- and eight-cylinder engines. The common practice presently employed is a *V* configuration such as in Figure 6.16.

Note the *V* configuration of the cylinders and pistons around the crankshaft. Such engines are popular because they provide a more balanced energy application to the crankshaft and are more compact. Six- and eight-cylinder engines are referred to as *V-6* and *V-8* engines, respectively. A current practice of the automobile industry is to place a *V-6* emblem on the hood or front fender of the vehicle to denote this engine type. In-line, six-cylinder engines are still quite common; whereas, in-line, eight-cylinder engines became extinct after World War II.

The four cycles of the gasoline engine are depicted in Figure 6.17. During the intake stroke, the piston is almost all the way down, the intake

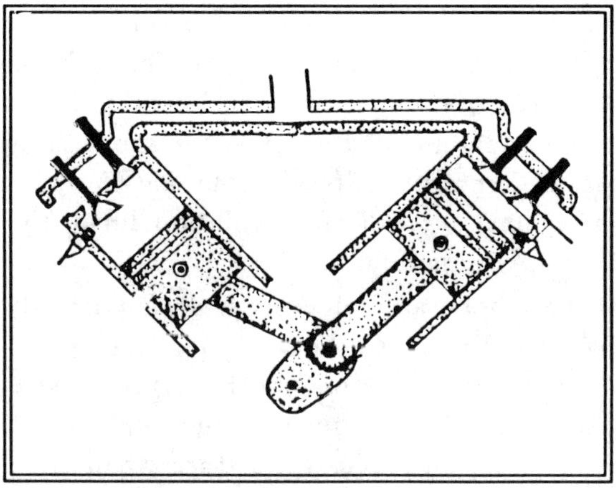

Fig. 6.16. A *V*-shaped cylinder configuration. Represented is a *V-2* engine.

valve is open and the exhaust valve is closed. The volume above the piston is increased and the pressure is reduced, allowing atmospheric pressure to force a fuel-air mixture into the cylinder. During the compression stroke (b), the intake valve is closed and the piston compresses the fuel-air mixture to its smallest volume. During the power stroke (c),

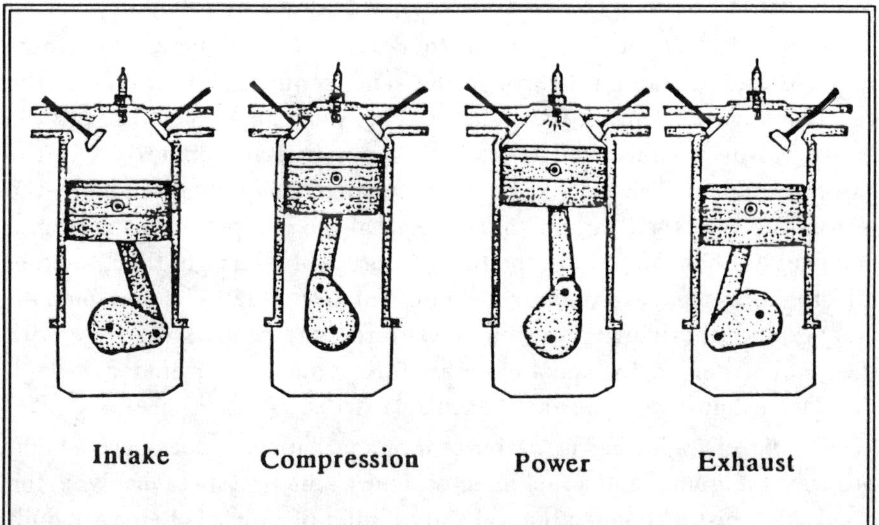

Fig. 6.17. Four strokes of a four-cycle gasoline engine.

both valves are closed, the spark plug ignites the fuel-air mixture and the piston is driven down. The crankshaft continues to rotate through the rest of the strokes due to the inertia of the flywheel. During the exhaust stroke (d), the exhaust valve opens and the piston forces the spent gases out of the cylinder. The process then repeats itself. Note that energy is applied to the crankshaft during only one half of a revolution during the two revolutions required to complete the four strokes.

The valves and electrical spark-generating device are driven directly from the crankshaft and carefully "timed" so the sequences take place at precisely the correct moment. The carburetor (not shown) is a device that works like a perfume atomizer. Air on its way to the intake valve picks up vaporized gasoline to form an explosive mixture when compressed in the cylinder.

The 4-cycle Diesel Engine

The diesel engine is a reciprocating, internal-combustion enginethat operates on the same principle as the gasoline engine. The fuel, however, is ignited by spraying it into the air in a cylinder that has been superheated by compression at the top of the compression stroke. A similar heat of compression can be felt when air is compressed in a bicycle pump.

Most diesel engines are of the four-stroke (cycle) type. The four-stroke diesel engine goes through the same sequence of strokes as the gasoline engine. During the intake stroke, pure air rushes through the intake valve into the cylinder rather than a fuel-air mixture as in the gasoline engine. The air is sealed in the cylinder and compressed by the piston to a very small volume at the peak of the compression stroke and becomes hot enough to burn the fuel. At precisely the right time, the fuel injector sprays in a pre-measured amount of diesel fuel. The fuel immediately explodes, driving the piston down during the power stroke. As with the gasoline engine, the spent gases are forced out of the cylinder through the open exhaust valve during the exhaust stroke.

The diesel engine has no carburetor or spark plug. It does have a high-pressure fuel pump and a fuel injector. Due to the pressures involved, the crankshaft, piston, connecting rod and cylinder of a diesel engine are built heavier than in a gasoline engine. Diesel engines are more efficient than gasoline engines.

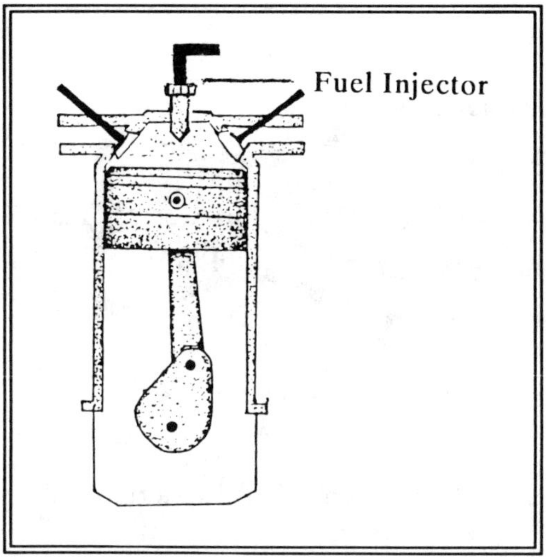

Fig. 6.18. A sketch of a four-cycle diesel engine.

The Gas Turbine Engine

The gas turbine is also an internal-combustion engine. It operates on the same principle as the steam turbine. The hot, expanding gases, which result from the combustion, transfer their energy to turbine wheels connected to the shaft (Fig. 6.19). As with the gasoline engine, the fuel-air mixture is compressed prior to ignition. However, in the gas turbine, the burning is continuous and the energy transaction is much smoother than in the reciprocating engine.

Air is drawn into one end of the turbine where it is compressed by air turbine wheels separated by stationery blades (stators) that direct the air at an appropriate angle. Fuel is picked up by the compressed air and is ignited. The burning gases expand against the turbine wheels, transferring their energy to the wheels. Eventually, the spent gases are exhausted. Electrical ignition is required only for starting the turbine because subsequent ignition is continuous.

Gas turbines are used in stationary electrical-generation plants and in turboprop airplanes. A turboprop airplane has a propeller as does the reciprocating airplane engine, but you can tell the difference between the two when the engines are turned off. A reciprocating propeller engine will quickly stop with a jerking motion as the pistons

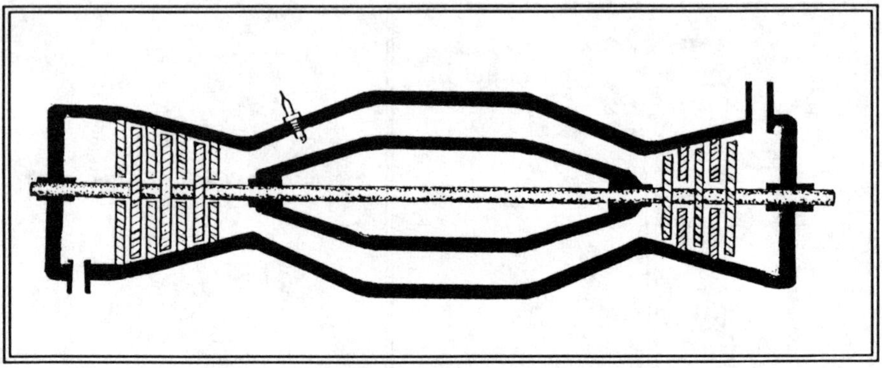

Fig. 6.19. Cross-section of the gas turbine.

work against the compression. A turboprop engine will gradually slow down to a smooth stop.

Jet Engines

Jet engines (Fig. 6.20) and gas turbines have many similarities. Both are internal combustion engines with a turbine air compressor and power turbine connected to a rotating shaft.

The major difference between the jet engine and the gas turbine engine is that the jet converts most of the energy of the burning gases to thrust, while the turbine transfers most of the energy of the burning gases into the rotation of the shaft. (The exception is the "fan jet" which converts "shaft energy" to thrust via the use of a fan.) Hot, expanding gases are forced out the back of the jet engine at a very high velocity. According to Newton's third law, a reaction force drives the engine forward in the opposite direction of the momentum of the gases. A small turbine wheel absorbs some of the hot gas energy to drive the air compressor.

Gas turbines and jet engines can be designed to be much more powerful for their physical size than reciprocating engines. This greater energy to mass ratio makes them desirable for large aircraft applications. Such engines are also much smoother than reciprocating engines. Remember, the piston in a reciprocating engine accelerates to full velocity, stops, reverses direction and accelerates to full velocity, again stops and reverses. Think of the effect on the pistons in an automobile engine as it reaches several thousand revolutions per minute while someone is drag racing. It is phenomenal that the engine holds together.

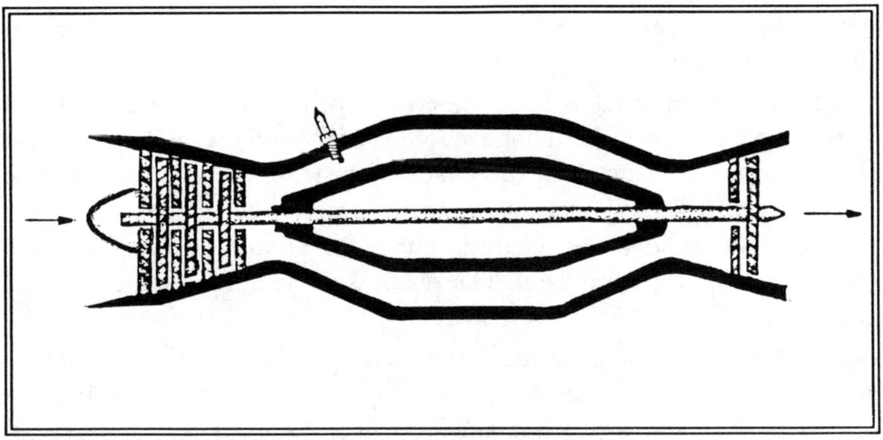

Fig. 6.20. Cross-section of a jet engine.

Science/Technology/Society

The use of heat can be a good example of the impact of science and technology on society. For example, the construction of bridges, large high-rise buildings, dams requires the use of long spans of materials that are often subjected to extreme temperature variations. Thermal linear expansion becomes a significant factor determining which materials will be used. Structural failure may result if inappropriate materials are used as a result of miscalculation or substitution for profit enhancement.

Heat is an important concept in peoples' interaction with the environment. Large population centers require huge quantities of energy for industrial processes, domestic heating and cooling, transportation and the generation of electricity.

Heat engines are an integral part of modern industrial-oriented life. A major portion of electrical energy generation and almost all present modes of transportation depend upon heat engines. Even electric trains get their energy from generators predominantly powered by heat engines.

A disadvantage of all heat engines is that they exhaust heat energy and pollutants into the environment. Some areas, such as Los Angeles, California, have a serious atmospheric pollution problem, due mainly to the exhaust of thousands of automobile engines. Jet aircraft contribute considerable amounts of hydrocarbon pollutants, as well as water vapor, to the atmosphere.

Stationary engines in large electrical power plants spew tons of water vapor and solids and huge amounts of wasted thermal energy into the atmosphere. Not all of the heat energy from power plants can be converted into useful electricity, so the waste heat must be ejected. Power plants tend to be located near large bodies of water because water has a high specific heat. The result is a rise in the temperature of the water.

What are some of the effects of warming the water in a bay, lake or river? In general, it changes the ecology. Plant and animal life are affected. The amount of one species (algae for example) is often increased while another (such as fish) may be decreased. What are other adverse effects? Beneficial effects? How are hydroponic techniques utilized to take advantage of waste heat?

What effect, if any, do the heat and water vapor rising from thermal power plants and the industries of a large city have on the atmosphere above that city? Can such heat and water vapor alter weather patterns? Contribute to smog formation? What are temperature inversions in the atmosphere?

Electrical companies in tropical climates such as in Hawaii have a major problem in efficient energy utilization due to heavy air conditioning demands during the daylight hours. Energy demands are minimal during the midnight to daylight hours, with the lowest demand at about 3:00 a.m. Consequently, electric-generating plants need to be built over-sized to handle peak energy demand periods. The problem of leveling out peak and minimal energy demand periods is partially solved by power plants using bio-masses such as sugar cane waste to generate electrical energy to add to peak demand periods. Another approach is to use the heat capacity of water as it changes from a solid to a liquid state. During minimal energy demand periods, large amounts of ice are frozen to utilize the energy available at that time.

During the heat of the day, the ice melts and is used to cool the building. Not only are electrical peaks leveled off, but a smaller air conditioning unit is required, thus lowering initial capital investment.

Solar energy is an alternate form of heating, both for hot water needs and for warming room air. The problem with the solar heating of room air is that the heat is only available, and in excessive quantities, when the sun shines. Certain solar heaters use the heat capacity of water to store the energy from the sun for later use. Short-wave solar radiation impinges upon water containers where it is transferred into long-wave radiation and raises the temperature of the water. The warm water or the air surrounding the water is distributed as required, especially during night hours or cloudy periods.

Noise is also a form of pollution. Psychological problems have been found to correlate positively with a high-noise environment. Hearing loss by those working around noisy machinery has become a serious concern. Jet aircraft can be especially troublesome. Living near a major airport is far from a pleasant experience.

It is becoming increasingly difficult to find a noise-free environment. Even if you find a moment of solitude, say on a Sunday morning, there is a good chance it will end quickly because of someone starting a lawnmower or revving up a motorbike. Noise from motorboats, chainsaws, dirt bikes and even farm tractors can intrude on beautiful lake and river settings. Even in remote wilderness areas, one is often bothered by the sound of aircraft, or in winter, a maverick snowmobile.

It does not seem probable that in the foreseeable future the air, water and noise pollution of the heat engine will be controlled. In a growth-oriented economy predicated on the concepts of more, faster and larger, it is, however, very important to ask serious questions about the implications of pollution and what is really meant by a "quality" existence.

Schools should teach students to examine individual actions. Value systems should be analyzed on the basis of discriminating between "wants" and "needs." Students must be taught to be good decision-makers.

Problems

1. The height of a copper spire on a church is 30 m. How much will the spire expand when the temperature changes from 15°C to 40°C? (coefficient of linear expansion for copper is 1.7×10^{-5}/°C)

2. A brass rod with a diameter of 2 mm is 20 cm long. A second brass rod is also 20 cm long, but has a diameter of 4 mm. If both rods are heated from 25°C to 35°C, how will the lengths compare at 35°C?

3. Why are periodic cuts made in the surface of a concrete highway?

4. A copper rod, 2 cm in diameter, expands 0.02 cm when heated to the boiling temperature of water. How much will a copper rod of twice the length of the first rod, and with a 4-cm diameter, expand when heated to the boiling temperature of water?

5. What is room temperature on the Celsius temperature scale?

6. What is the freezing temperature of water on the Celsius temperature scale?

7. Rod A is 50 cm long and has a diameter of 0.3 cm. Rod B is 100 cm long and has a diameter of 0.6 cm. If the two rods are heated from 25°C to 95°C, and rod B becomes 100.084 cm long, what is the length of rod A?

8. What is the boiling temperature on the Celsius scale of water at sea level?

9. An iron rod 10 m long expands 0.0012 cm when heated to a certain temperature. How many centimeters will another iron rod 10 m long and of twice the diameter as the first rod expand when heated through the same temperature range?

10. If a volume of hydrogen gas is heated to 98 °C, its volume becomes 150 cm^3. If an equal volume of oxygen is heated to 98 °C, what will be its volume?

11. Is the volume expansion of a liquid a characteristic property?

12. How many K-Cal are required to change the temperature of 100-kg of water from 50 °C to 90 °C?

13. If the temperature of 50 kg of water is changed from 80 °C to 20 °C, how many K-cals of heat have been removed?

14. If you eat a dessert which has a heat equivalent of 1500 K-Cals, how many gallons of ice water would you need to drink to neutralize the K-cals ingested?

15. The temperature of a certain liquid is observed at one-minute intervals and recorded as follows:

min	°C	min	°C
0	86	5	78
1	84	6	79
2	81	7	77
3	79	9	75
4	79	10	73

What is the freezing/melting temperature of the liquid?

16. Which is a more efficient engine, the steam turbine or the reciprocating steam engine?

17. Which is the more efficient engine, the reciprocating gasoline engine or the gas turbine? Which is a smoother-running engine?

18. How many cylinders does a four-cycle gasoline engine have?

19. In what cycle is the piston of a four-cycle, gasoline engine if the exhaust valve is open and the intake valve is closed?

20. In what cycle is the piston, if the exhaust valve and the intake valve are closed in a four-cycle gasoline piston engine?

21. In what cycle is the piston of a reciprocating gasoline engine if the intake valve is open and the exhaust valve is closed?

22. In what cycle is the piston of a diesel engine if the intake valve is open and the exhaust is closed?

23. What ignites the fuel-air mixture in a diesel engine during the power stroke?

24. In what ways do a diesel and a reciprocating gasoline engine differ?

25. Is noise a form of pollution? Name some noises which you feel are forms of pollution.

Chapter 7

Solutions

Thus far, you have examined several characteristic properties used to identify substances. However, two or more substances may have one or more identical properties, making positive identifications difficult. Therefore, more than one property needs to be studied to ensure accuracy. As a case in point, the hydrocarbons cycloheptane, n-butanol and sec-butanol each have a density of about 0.81 g/cm^3. Therefore, you need to look at one or more other characteristic properties to distinguish one from the others. Checking the boiling temperature (118°C for cycloheptane, 118°C for n-butanol and 100°C for sec-butanol) allows you to distinguish between sec-butanol and the other two. But to distinguish between cycloheptane and n-butanol, you need to examine another property, such as their melting temperatures (–90°C and –89°C, respectively).

Even when a characteristic property is quite different for two substances, a researcher will usually examine several properties in order to feel secure in discrimination and identification. Voluminous handbooks of the properties of substances have been published as "standards" to aid the researcher in making positive identifications of materials. In this chapter you will be looking at another characteristic property that will help you identify and discriminate between materials.

*****

Experiment 7–1. Solubility of Solids

Perhaps you have noticed some people adding large amounts of sugar to their coffee. After drinking the coffee, a considerable amount of sugar remains undissolved at the bottom of the cup. The coffee was "saturated" with sugar – it could hold no more in solution. The addition of more sugar than the coffee can dissolve apparently gives only psychological satisfaction to the coffee drinker.

In this experiment you will attempt to quantify solubility to determine if it is a characteristic property. Half of the class will add approximately

5 g of NaCl to test tubes containing about 5 g of water; the other half will add 5 g of $NaNO_3$. Place a stopper in your test tube and shake it until you feel the maximum amount of solid has dissolved (Do not exceed 5 minutes).

Next, determine the mass of a clean, dry evaporating dish on an equal arm balance and record. Carefully decant (pour off) as much of the liquid as you can from the test tube into the evaporating dish, allowing no undissolved solid in the evaporating dish. Be sure to grasp the test tube near the top with your thumb, index finger and middle finger so as not to warm the test tube.

Gently heat the evaporating dish until all the liquid has evaporated. To reduce spattering, you should reduce the heat by moving the burner away from the dish in progressive steps as soon as a solid appears (Fig. 7.1).

Determine the mass of the evaporating dish and the dry solid using the double pan balance and record the value.

Fig. 7.1. Evaporating the water from the solution.

Solubility is conventionally expressed as the ratio of the mass of the dissolved solid to 100 cm^3 of water. You must first determine the amount of dry solid per each gram of water by:

$$\frac{g\text{ - }solid}{g\text{ - }water}$$

where:

$$g\text{-}solid = g\text{-}dish + g\text{-}solid - g\text{-}dish$$

and:

$$g\text{-}water = (g\text{-}dish + solution) - g\text{-}dish - g\text{-}solid$$

Because 1 *g* water has a volume of 1 cm^3, the relationship can be expressed in g/cm^3 water. Multiplying both the numerator and the denominator by 100 gives the solubility in the conventional terms of 100 g/cm^3 water.

Compare your results with your classmates' findings. Does it appear that the solubility is different for the dissolved solids? Is solubility a characteristic property?

*****

Experiment 7–2. Temperature as a Function of Solubility

In the previous experiment no consideration was given to temperature, but you may have observed that your test tube felt cooler as the dissolving progressed. To determine whether temperature is a factor in solubility, add about 10 g of a solid to 10 cm^3 of water in a test tube. Different kinds of solids will be distributed to members of the class to determine whether the concept applies to more than one substance.

Mount the test tube in a beaker one-third full of water as in Figure 7.2. Heat the water in the beaker while stirring the water inside the test tube, using a thermometer as a stirrer.

CAUTION: Thermometers are fragile and they contain mercury, which gives off hazardous vapors if spilled.

Record the temperature when no more solid will dissolve by further heating and stirring. Do not boil the water inside the test tube.

Compare the results with those obtained by your classmates. Did all of the solid dissolve as the water was raised to its boiling temperature?

Does temperature affect solubility? Does temperature affect the solubility of all substances the same?

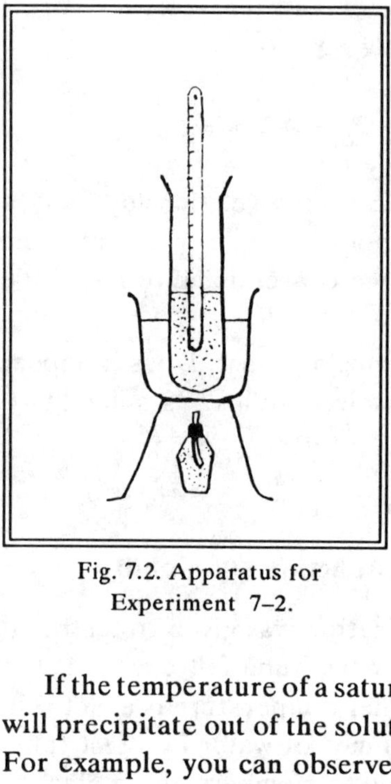

Fig. 7.2. Apparatus for Experiment 7–2.

*****

The solubilities of some common substances are shown in Table 7.1. Figure 7.3 is a graphic representation of the data. You can read the solubility of any of the substances at any temperature by following the vertical line from a temperature value until it intersects the curve of the substance of interest. Then, by following the horizontal line from that point to the left, you can read the solubility in grams of solid per 100 cm^3 of water. Remember that any point on the curve for a substance represents the temperature at which the water solution of that substance is saturated.

If the temperature of a saturated solution is lowered, the dissolved solid will precipitate out of the solution until the solution is again at saturation. For example, you can observe on the graph that a solution of potassium nitrate is saturated with 120 g/100 cm^3 of water at 65^0C. If the temperature of the solution is lowered to 10°C, the solution can hold only 22 g/100 cm^3 of water. Therefore, 120 g – 22 g = 98 g of potassium nitrate will precipitate out.

Conversely, if the temperature is held constant but part of the water is evaporated, the amount of solid it can hold changes. If you have 100 cm^3 of water with 118 g of sodium nitrate dissolved in it at 55°C, it is saturated. Suppose you boil away 50 cm^3 of water. How much sodium nitrate will precipitate to the bottom of the container? According to the graph, 100 cm^3 of water is saturated with 118 g of sodium nitrate at 55 °C, so 50 cm^3, or one-half of 100 cm^3, will be saturated with one-half of 118 g of sodium nitrate or 59 g. Therefore, 59 grams of sodium nitrate will precipitate out.

Table 7.1. Solubilities of some substances in *g-solute* per 100 cm^3 H_2O.

°C	Sodium Nitrate	Potassium Nitrate	Sodium Chloride	Potassium Chlorate
0	73.0	13.9	35.7	3.3
10	80.8	21.2	35.8	5.2
20	87.6	31.6	35.9	7.3
30	94.9	45.3	36.1	10.1
40	102.0	61.3	36.4	13.9
60	122.0	106.0	37.1	23.9
80	148.0	167.0	38.0	37.6
90	162.0	203.0	38.5	46.0
100	180.0	245.0	39.2	56.3

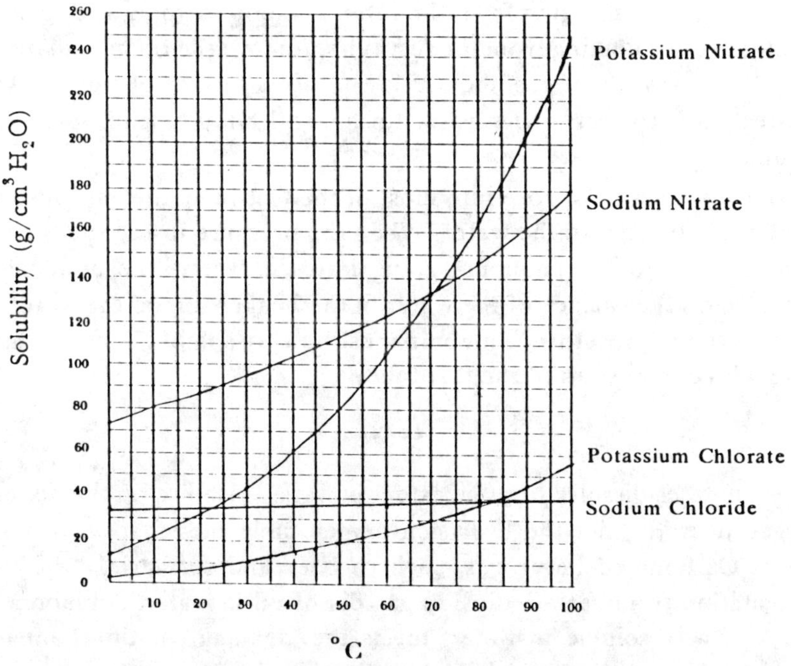

Fig. 7.3. Graphic representation of data in Table 7.1.

Solubility as a function of temperature is a characteristic property of a substance. One can readily see the utility of having a temperature/solubility graph such as Figure 7.3. Remember, however, that identifying a substance by one characteristic property is not fool-proof. For example, the solubility of, sodium chloride and potassium nitrate are about the same at 23°C. At 72°C the solubilities of potassium nitrate and sodium nitrate are the same, and at 83°C the solubilities of sodium chloride and potassium chlorate are the same. Consequently, you cannot feel secure in basing conclusions on only one piece of data.

*****

Experiment 7–3. The Solubility of a Gas

Place about 10 cm^3 of ammonia water in a small test tube with 5–6 boiling chips. Insert a one-hole stopper, in which a 12-cm piece of glass tubing has been inserted, so that it just protrudes through the stopper (Fig. 7.4). Invert a test tube over the open end of the glass tubing and collect any gas emerging from the tube. As soon as you notice a strong ammonia odor coming from the test tube, remove the flame and quickly move the test tube of gas, mouth down, into a beaker of water. Allow the test tube to rest in the water for several minutes and observe the results.

Because the water rose into most of the test tube, the gas must have readily dissolved into the water. When the pressure inside the tube was reduced due to the ammonia going into the water, the atmospheric pressure on the surface of the water in the beaker forced the water into the test tube. Therefore, ammonia must be very soluble in water, as the water filled almost all of the test tube.

*****

Most gases dissolve to some extent in water. Fish live on the dissolved oxygen in water, not the hydrogen-oxygen molecules that make up the water. Carbonated beverages, such as beer and soft drinks, get their carbonation from carbon dioxide gas dissolved in water. Ammonia gas, which is very soluble in water, forms the cleaning solution "ammonia water" or "aqueous ammonia" when dissolved in water.

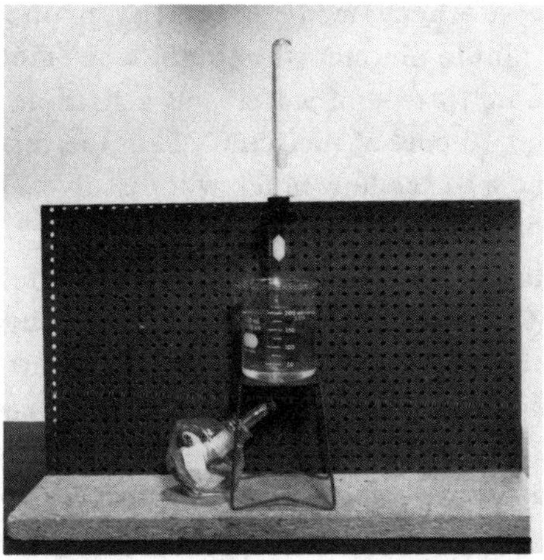

Fig. 7.4. Apparatus for generating ammonia gas.

The solubility of gases is inversely proportional to the temperature and directly proportional to the pressure. If you pour cold soda beverage into a glass, foam will form on the top and gas bubbles will be present throughout the liquid. As the soda warms, more gas bubbles will move to the top, and eventually, as the liquid reaches room temperature, the gas will stop forming. The soda will then taste very flat due to the absence of the gas. As the pressure is released and the temperature rises, water can hold less gas. For example, as the temperature of water in a pond rises, less oxygen remains dissolved for fish to live on. Even a glass of tap water, normally below room temperature, will form bubbles of dissolved air as it warms to room temperature in the glass.

*****

Experiment 7–4. Alcohols as Solvents

Alcohols are solvents for certain materials. For example, many people believe that iodine is a liquid. Actually, it is a solid that is routinely marketed for medicinal purposes as "tincture of iodine," an alcohol solution. Ethanol, also called *ethyl alcohol* or *grain alcohol,* is found in alcoholic beverages and is used to dissolve food flavorings such as vanillin, mint, and almond oil.

As an example of the varying solubility properties of certain materials, place a minute amount of naphthalene (moth balls) in a test tube containing about 10 cm^3 of water and a similiar amount in a test tube containing abut 10 cm^3 of methanol. Seal the test tubes with your thumb and shake them to see how much will dissolve. Does naphthalene appear to dissolve in water? In alcohol?

Now add some tap water to the test tube with the methanol. What do you observe? What caused this? Place a pinch of sugar in a test tube containing 10 cm^3 of water and a test tube containing 10 cm^3 of methanol. Shake each test tube and observe what happens. What do you predict will happen if you add some methanol to the sugar and water? What was the effect? Did you get the effect you expected? Why or why not?

Although water is known as the universal solvent, it is by no means the only solvent. Petroleum-based solvents are used to clean machinery parts and painting equipment. So-called dry cleaning fluids are solvents that dissolve the grime from clothing. Such solvents are liquids at room temperature, but because they do not contain water and evaporate from clothing very rapidly, they are considered "dry cleaning" solvents.

*****

Science/Technology/Society

The use of certain solutions is central to an industrial society. Domestic tasks such as washing clothing and dishes involve dissolving oils and forming water suspensions. The entire soda beverage industry is based upon dissolving carbon dioxide in water through the use of pressure. Many forms of marine life depend upon the dissolved oxygen in water for life, and, therefore, the fishing industry indirectly depends on a dissolved gas.

It appears that nature maintains the environment in a fairly precise balance. A certain amount of dissolved oxygen is needed to sustain specific animal life in water, but if too much oxygen is dissolved, some vegetable life, such as algae, will reproduce at too fast a rate. If this causes an over-concentration of certain gases, the pH of the water is altered to the detriment of aquatic life.

Farm fertilizers, pesticides and herbicides placed on the land are dissolved by atmospheric precipitation and their water solutions washed into surface water. Such solutions also percolate through the subsoil into the water tables, which are sources of well water. Serious concerns are being raised about the quality of well water in agricultural areas, due to the increasing percentage of nitrates and toxins. Shallow wells on the farms may no longer be desirable.

Expelling industrial solvents into the atmosphere can be harmful to vegetation, and if the solvents are dissolved in rainwater, soils and water supplies may be adversely affected. Acid rain, the result of sulfides, sulfates, carbon monoxide and carbon dioxide dissolving in atmospheric precipitation, is responsible for killing entire forests in North America and Europe. Such pollution harms vegetation around the source of pollution, and prevailing winds can carry them to distant areas.

It is clear that many solutions, such as beverages, medicines and cleaning fluids, can be beneficial to mankind. However, liquids, solids and gases can form solutions which adversely affect the environment, and care must be exercised in their use and disposal. The possible side effects of the solutions used by industry, farmers and homemakers should be studied to determine necessary precautions.

Problems

1. How many grams of sodium nitrate can 100 cm^3 of water hold at 70°C?
2. A mass of 20 g of sodium nitrate is dissolved in 20 cm^3 of water. At what temperature would you first observe a precipitate as the solution is cooled?
3. How many grams of potassium chlorate can 50 cm^3 of water hold at 60°C?
4. Fifty grams of potassium nitrate are dissolved in 100 g of water at 90°C. As the solution is cooled, at what temperature would a precipitate form?
5. Sixty grams of potassium nitrate are dissolved in 150 cm^3 of water at 90°C. If 50 cm^3 of water are boiled away and the

solution is allowed to cool, at what temperature will a precipitate form?

Chapter 8

Some Methods of Separation

Thus far, you have used the knowledge of characteristic properties to identify substances. But knowledge of characteristic properties can also be used for other purposes. One of the primary ways such information is used is in the separation of substances for research or industrial purposes. Whole industries have been developed around separating useful materials from naturally-occuring mixtures on the basis of different characteristic properties. For example, some substances are soluble in some liquids but not in others. So one component of a mixture may be dissolved in a specific liquid and the insoluble materials filtered out. A mixture of liquids can be separated by raising the temperature of the mixture and collecting each constituent at its respective boiling temperature. Materials also have unique freezing temperatures and constituents of a mixture of liquids freeze or crystalize at specific temperatures, allowing the solids to be removed at their respective freezing temperatures. Other methods of separation involve capitalizing on differing densities and diffusion rates. In this chapter you will examine a variety of these processes.

*****

Experiment 8–1. Separation of Mixtures and Solids

Given a mixture of iron filings and salt, you could separate the two with a magnifying glass and a set of tweezers, a tedious chore indeed. An easier and faster method would be to use a powerful magnet to draw out the iron filings. Can you suggest another approach?

The following experiment may give you a clue. You will work with a mixture containing sulfur and sodium chloride particles. Pour about 3 g of the mixture into 10 cm^3 of water in a test tube. Place a stopper in the test tube and shake it for a few minutes.

Fold a circular piece of filter paper in half and the remaining portion in half again. Spread an outside, single layer of the triangle formed from the triple-layered side, form a cone, and place it in a funnel set in a test tube (Fig. 8.1). After filtering all the liquid through, flush the filter paper by adding 5–10 cm^3 of water. Catch the liquid in the test tube. What do you observe in the filter paper?

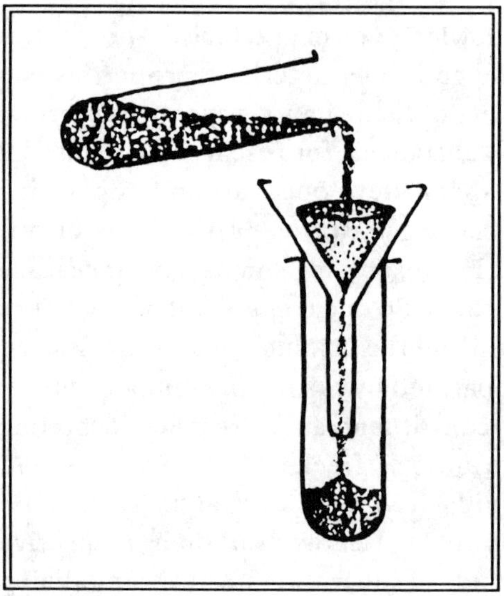

Fig. 8.1. Apparatus for Experiment 8–1.

Place a small amount of the filtered liquid (*filtrate*) in an evaporating dish and evaporate to dryness with an alcohol burner. Is anything left behind? What does it look like? Were you able to separate the two substances?

You used the solubility in water characteristic of the two substances to separate them. One material was not soluble in water and was held back by the filter paper. The particles of the second substance could not pass through the filter paper when dry, but when in solution, became small enough to easily pass through. Is it possible that the dissolving process breaks up the particles into nearly atomic sizes?

*****

Experiment 8–2. Fractional Distillation

Place 20 cm^3 of an alcohol-water mixture into the test tube of the apparatus shown in Figure 8.2. Heat the hot water bath and record the temperature of the alcohol-water mixture every minute. You should continue the process until you observe the temperature abruptly increase and then become constant. Quickly replace the collection test tube with a clean test tube. Take several more readings and remove the rubber tube from the collection test tube and extinguish the burner.

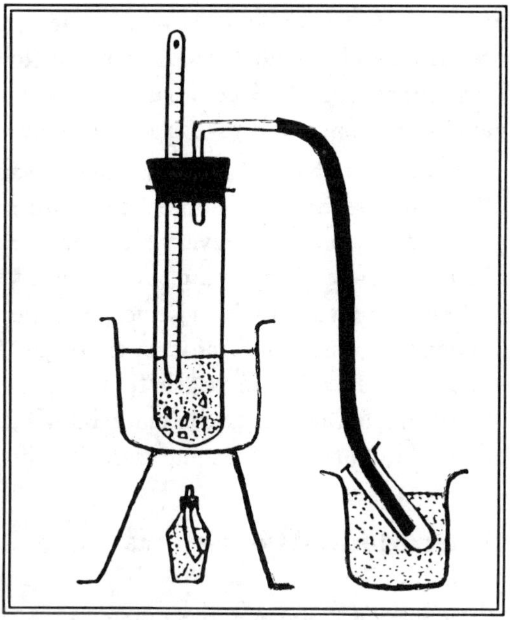

Fig. 8.2. Apparatus for Experiment 8–2.

CAUTION: Do not allow the end of the rubber tube near the flame as alcohol is very flammable!

Carefully sniff the liquids you have collected in the two test tubes. Can you determine what the liquids are? Graph the data you have collected. Can you detect the boiling temperature of the alcohol and the water? Can you determine which alcohol you have by its boiling temperature? At what times would it be best to collect the purest alcohol? At what time would you begin to collect the purest water?

*****

Separating liquids by heating and then condensing them at their boiling temperatures is known as *fractional distillation*. "Fractions" of the original mixture are collected.

Ethanol is routinely fractioncd out of a mixture of ethanol and water. Water and ethanol are the products of grain fermentation. The mixture is maintained at the boiling temperature of the ethanol, which is then condensed and collected. If a higher purity is desired, the collected material can be re-distilled. A more efficient method that yields a highly pure substance is to construct the distilling apparatus so the alcohol vapors condense and re-evaporate in successive steps.

Fractional distillation is also used to separate the components of crude oil, a mixture of many substances, into several constituents. The crude petroleum is heated to a very high temperature and injected into the base of a *meta tower* (fractioning tower). The vapors condense at or near their boiling temperatures in a series of pans inside the tower. The liquids condense as discrete liquids, such as gasoline of different octane ratings, kerosene, diesel fuel, paraffins and asphalt. The lower the boiling temperature, the higher the material will rise before condensing. Asphalt and tars tend to condense near the bottom.

Other Methods of Separating Substances

In certain instances, too much heat is detrimental to the desired end-product, so other methods must be employed to achieve purity. Solids that crystallize from a solution have a higher purity of the solid than has the mixture. Dissolving the crystals in pure water and recrystallizing will increase the purity of the solid. The process is repeated until the desired level of purity is reached. This process finds application in the pharmaceutical industry.

Air is a mixture of gases. Individual gases comprising air can be separated by supercooling the air to the respective boiling temperatures of the constituent gases to achieve liquification and separation. Each liquid constituent can be stored and warmed to above its boiling temperature to convert it back to a gas, such as the case with liquid nitrogen. Carbon dioxide (CO_2), when cooled, forms a solid called *dry ice*, which sublimes into a gas when warmed.

*****

Experiment 8.3. Paper Chromatography

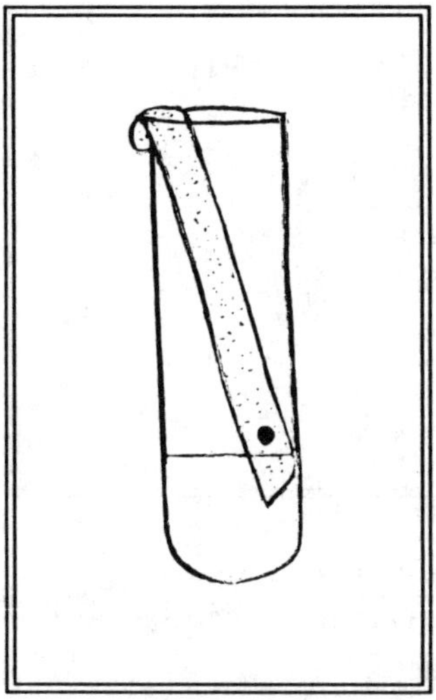

Fig. 8.3. Paper chromatography detail.

Pour water into a large test tube to a level of about three centimeters. Cut a strip of filter paper to a width that will conveniently fit into the length of a test tube. Place a dot of ink about one centimeter from the end of the paper and suspend the strip with the dot end of the paper just touching the water. Other members of the class may use different colored inks or food coloring. Be sure you keep the ink spot above the water level (Fig. 8.3). Allow the apparatus to stand for 5–10 minutes. Observe the separation of the ink spot. Compare it with other materials used by your classmates.

The ink particles move along with the water as it diffuses through the paper by its molecular motion. The rate of movement is dependent on the adsorption rate of the particles by the filter paper. The less dense particles are carried further up the paper than the more dense particles. You can distinguish the particles by their respective colors. You have, in effect, separated the constituent colors of the ink. Because the movement of the particles is dependent upon their adsorption rates, particles found in one substance will move to the same zone in a piece of filter paper as like particles found in another substance moving through another piece of filter paper (Fig. 8.4).

Once the substances have been identified, their relative positions can lead to their identification in other materials.

*****

Science/Technology/Society

Prior to the twentieth century, there were only limited uses for petroleum. The popularity of the gasoline engine in such devices as the

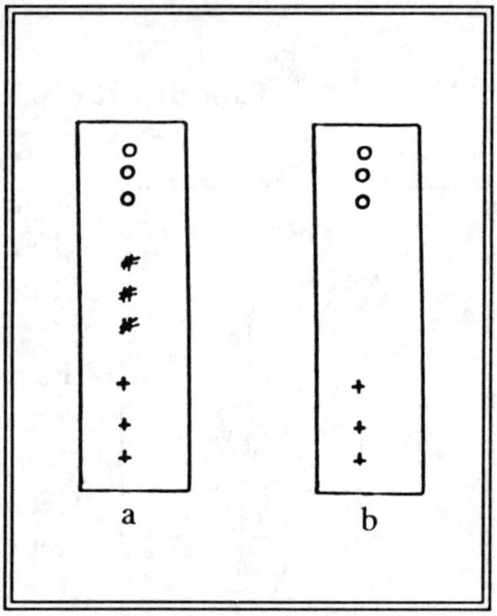

Fig. 8.4. Three colors are found in substance a, two in substance b.

automobile and the airplane brought increasing demands for high-grade, high-test gasoline. But its refinement through fractional distillation led to many unutilized components and a lot of waste.

Science and technology have advanced such methods as "cracking" large molecules of petroleum, which could not be utilized by fractional distillation and were formerly wasted, into smaller molecules to produce more gasoline. Technologies are now in use that make more efficient use of the resources available, helping stretch precious energy pools.

The production of renewable or synthetic energy resources is possible through various separation processes such as distillation. Ethanol can be recovered from the fractional distillation of sugar fermentation. Petroleum products such as gasoline, as well as ethanol itself, can now be derived from coal.

In this chapter paper chromatography was added to your repertoire of methods for identifying substances. Materials appearing in very small amounts can be identified using this process. Impurities in the environment often need to be identified so the source may be found or dangers caught in time. Such processes make the task possible.

Chapter 9

Atoms, Elements, Ions and Compounds

Up to this point the discussion and activities have been concerned with physical changes such as condensation, evaporation, boiling and freezing, and linear and volume expansion. The focus was on changing substances in some manner and then reversing the process to restore them to their original form.

You will now deal with the fundamental building blocks of nature. The preceding discussion alluded to a science based on the particle nature of matter. Scientists' interests are in the identification of the basic particles of nature and a model of how they combine, separate and influence each other.

For centuries, people have contemplated the make-up of matter. Various theories and categorization schemes have been suggested, only to be discarded later. But slowly and steadily a theory has developed and continues to evolve.

Early Greek thinkers formulated the basis for the present theory on the structure of matter. Around 580 B.C. Thales postulated that all matter consisted basically of water. He viewed water as a primary element. During the fifth century B.C., Leucippus, a Greek philosopher, formulated the concept that all matter is composed of small, indivisible particles. His student, Democritus, named the basic particles "atoms." The great philosopher Aristotle (384–322 B.C.) concluded that the earth was made of only four basic elements, earth, fire, air and water, and that the heavens were made of a fifth element, "aether." In about the year 50 B.C., Lucretius, a Roman philosopher and poet, summarized the fundamental Greek philosophy of atomism in a poem entitled "On the Nature of Things." The Greek theories were strictly philosophical, authoritarian dogmatisms with no basis in experimentation.

The Middle Ages were a time of scientific stagnation dominated by the philosophies of Aristotle. Medieval religious dogma controlled intellectual pursuits. Economic policy under the agrarian-based feudal system was not conducive to technical improvements and little was done to advance scientific knowledge. The study of matter was usually done by alchemists, whose work was often based on mysticism and whose motivation was to convert cheap materials into precious metals.

A revival in scientific interest, called the *Medieval Renaissance,* began during the latter part of the twelfth century and continued into the thirteenth century. By the middle of the seventeenth century, some alchemists had suggested mercury, sulfur and salt as basic elements.

By the latter part of the seventeenth century, philosophic authority in science gave way to empirical experimentation. In 1789 Atoine Lavoisier, in his *Elements of Chemistry*, described "elements" as substances not subject to further chemical decomposition and introduced the principle of *the conservation of matter*—no material is gained or lost during a chemical reaction. Louis Proust (1755–1826) observed that elements combine in definite proportions to form compounds. His idea was confirmed experimentally by Jean Servais Stas about fifty years later. Chemistry was at last moving from alchemy into empirical experimentation.

John Dalton, during the nineteenth century, worked with gases in his search for the structure of matter. He suggested an atomic theory based on the idea of a unique atom for each element. His work on the combining ratio of elements to form compounds laid the foundation for modern atomic theory, although his ratios were later shown to be in error.

These early chemists were able to accumulate an impressive amount of information on the structure of matter through chemical analysis and synthesis. However, it was theoretical physics that paved the way for present atomic theories. The discovery of X-rays and radioactivity and the development of high-energy particle accelerators made it possible to probe indirectly into the interior of the atom, thus, leading to the present model of the structure of matter.

Present-day science is predicated on the concept of the particle theory of matter. Although no one has been able to see an atom directly, a host of information has been amassed through such indirect processes as X-ray crystallography, light spectroscopy, mass spectroscopy, electron microscopy and particle acceleration.

Much of what has been learned through the physical sciences has been discovered through indirect means – not through direct observation. If you have a problem conceptualizing a model of things not observed, consider gravity. You would be hard-pressed to draw a picture of gravity; yet, a considerable amount is known about the effects of gravity, and mathematical relationships can be set up to predict the path of bodies in a gravitational field. Likewise, electricity is a construct never seen. Charges cannot be observed moving in a conductor, but you can feel an electrical shock or observe a glowing light bulb. Electricity is used routinely by many who know little, if anything about it. Have you ever seen air? What evidence do you have that air exists?

Elements

Certain naturally-occurring substances appear homogeneous; whereas others are obviously heterogeneous. A piece of granite appears to be composed of different colored particles. Crushing the granite yields pieces of feldspar, mica and quartz. A cross-section of a piece of wood appears to show variance in composition. Therefore, granite and wood are not considered basic elements because they can be broken down into simpler substances.

A piece of native copper (that which occurs naturally in the metallic state), on the other hand, appears homogeneous when sawed into two equal-size pieces. Cutting it into progressively smaller pieces and inspecting them with a magnifying glass continues to reveal only copper. It would appear, then, that copper cannot be broken into simpler substances and might be a basic element. Tests that subject copper to various chemical analyses reveal that, although copper will combine with a host of other substances, it cannot be broken down into simpler substances. Therefore, copper is considered a basic element.

A group of metals known as *alloys* also appear homogeneous throughout. Upon melting them, however, one can observe a separation of more than one substance. When the alloys are treated with chemicals, such as acids, certain constituents of the metal may react while others do not. Alloys are, therefore, mixtures of metals and are not basic elements.

A very large class of materials, known as *compounds*, are also homogenous in constituency. For example, a pure crystal of common table salt (NaCl) can be divided into smaller and smaller pieces, each with a homogenous constituency. A critical point can be reached where dividing the salt crystal one more time yields two particles that are no longer salt. One is sodium, a metal which reacts spontaneously with water; the other is chlorine, a poisonous gas! These two particles in no way resemble the original harmless-appearing piece of table salt, but neither can be broken down by chemical or physical means. Therefore, table salt is not a basic element but consists of sodium and chlorine, which are basic elements.

An element is the simplest division of a homogeneous substance that cannot be decomposed by chemical means and retains the identity of that substance. Common examples of elements are iron, copper, hydrogen, oxygen, aluminum, nitrogen, lead, mercury and nickel. A total of 104 elements have been identified. Each element is given a specific number according to a chemical characteristic scheme. You will study this scheme in the Periodic Table of Elements section of this chapter. Some of the elements with high numbers are very unstable and have been produced only in the laboratory, existing for only short periods of time.

It is not possible to see an atom with the naked eye or even through the most powerful optical microscope. All of the information upon which the present atomic model has been derived has been acquired by indirect means. Electron micrographs of a very fine object, such as the tip of a needle, have been used to produce photographic images of the atoms at the very tip of the needle. Viewing these images, which are produced on the screen of a cathode ray tube, is comparable to seeing the president of the United States on television. You recognize the president as a familiar face, even if you have never seen him/her in person.

Ingenious devices have been invented that use high-energy particles and X-rays to decompose basic atoms to determine their components. It turns out that no particles have been identified that are unique to a specific element. All elements are made of one or more of the basic particles that appear to be the basic "building blocks" in nature–the material of which all matter is made.

The predominant mass of atoms appears to be concentrated in a very dense, positively-charged, central *nucleus.* Surrounding the nucleus is a "cloud" of negatively-charged particles (*electrons*), with a total charge equal to the total positive charge of the nucleus, resulting in a net charge of zero. All atoms have a neutral, or zero, net charge. On an atomic scale, the overall electron cloud is located at a considerable distance from the nucleus. Therefore, the volume of an atom is predominantly empty space.

Electrons follow discrete patterns (*energy levels*) located, on the average, at definite distances from the positively-charged nucleus. Analysis of experimental evidence suggests that the electrons surrounding the nucleus of an atom are constantly in motion. They are held in place by the centripetal acceleration caused by the attraction of the oppositely charged nucleus. However, as you will observe in Chapter 12, charges in motion radiate energy. So the moving electrons *should* consistently lose energy and move constantly closer to the nucleus, eventually collapsing into the nucleus. This, however, *is not* consistent with experimental observation. According to wave mechanics, the electrons in motion follow wave motion behavior as well as particle characteristics. The electrons can possess only discrete energies, *quanta*, that are determined by the frequency of their vibration. They set up standing waves (Chapter 10), and, therefore, do not radiate energy.

The location of an electron within an energy level at any given time is probabilistic. This means that its location at any one instant of time cannot be predicted but is a "chance event," such as the toss of a coin. Mathematical equations can estimate the location and position of the electrons, but it is difficult to achieve a visual representation .

Major electron energy levels that represent discrete amounts of energy (*quanta*) are designated with *quantum numbers* ($n = 1, 2, --- 7$). Beginning with $n = 1$ closest to the nucleus, the energy level increases as the quantum number increases. An atom is in a "stable condition" when electrons assume energy levels characteristic of that atom.

If energy is added to an atom (e.g., by being "excited" by a high electrical potential), an electron may accept only discrete amounts of energy that will raise it to a higher energy level. When the electron loses this "excited" energy, it returns to its normal energy level. The lost energy is radiated in the form of light. Energy can be added to excite electrons of atoms of a gas, as in neon tubes or fluorescent lights. As the electrons return to their original energy levels, they give off photons (packets) of light energy.

Not only do electrons move around the nucleus in distinct energy levels, but, as evidenced by applying magnetic fields, possess individual spins in specific directions. Each individual spin is counteracted by the opposite spin of an electron with which it pairs in a suborbit or *orbital.* Orbitals, by convention, are designated by the letters *s, p, d* and *f*, with each forming a unique pattern. The *s*-orbital assumes a basic spherical pattern, but the remaining patterns require a study of wave mechanics that is beyond the scope of this text.

Quantum levels can incorporate only n^2 orbitals each, and, therefore, levels $n = 1, 2, 3$ and 4 can incorporate only 1, 4, 9 and 16 orbitals, respectively. Each orbital represents a maximum of two electrons; therefore, the maximum electron capacity for quantum levels $n = 1, 2, 3$ and 4, is $2n^2$ or 2, 8, 18 and 32, respectively. Atoms are most stable when their outermost quantum level is filled to its maximum capacity.

Atoms are routinely represented as concentric circles of electrons located at various distances from the nucleus, similar to the planets rotating around the sun in the solar system. This design is more for convenience than accuracy. A more realistic representation shows a densely-packed nucleus, with a somewhat vague boundary, surrounded by a nondescript cloud of electrons, with varying concentric concentrations at different distances from the nucleus representing energy levels (Fig. 9.1). For convenience, the conventional planetary model will be used in the following discussions, but you should be aware that these are not representations of how an atom actually appears.

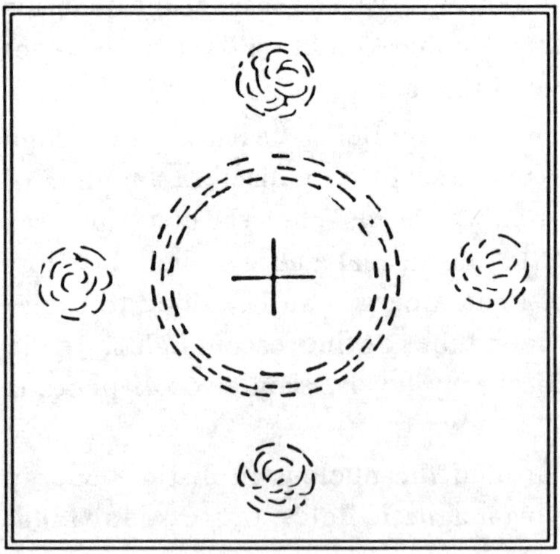

Fig. 9.1. Representation of an atom.

Figure 9.2 represents the atomic configuration of the elements hydrogen, helium and neon. Hydrogen, the simplest of the elements, has a nucleus of one positively-charged proton and one electron orbiting about it in quantum level $n = 1$. Helium has a nucleus consisting of two protons and two neutrons. A *neutron* is a neutrally-charged particle with a mass very close to the mass of the proton and is involved in holding the nucleus together. The first quantum level of helium, which consists of one orbital, is filled with two electrons. Neon has a nucleus of 10 protons and 10 neutrons. Notice the first quantum level is saturated with one orbital (2 electron) and its second quantum level is completed with four orbitals (8 electrons). Both helium and neon are very stable gases, with their highest quantum levels complete. Chemical reactions transpire between the unstable (unfilled), outer quantum levels of atoms, and therefore, helium and neon tend not to react with other elements to form compounds, although helium has been shown to form compounds under very special conditions.

The number of protons in the nucleus is the *atomic number* of the atom. In Figure 9.2 one observes the atomic number of hydrogen as one (1), helium as two (2) and neon as ten (10). In each case the number of protons corresponds to the number of electrons, making the net electrical charge zero.

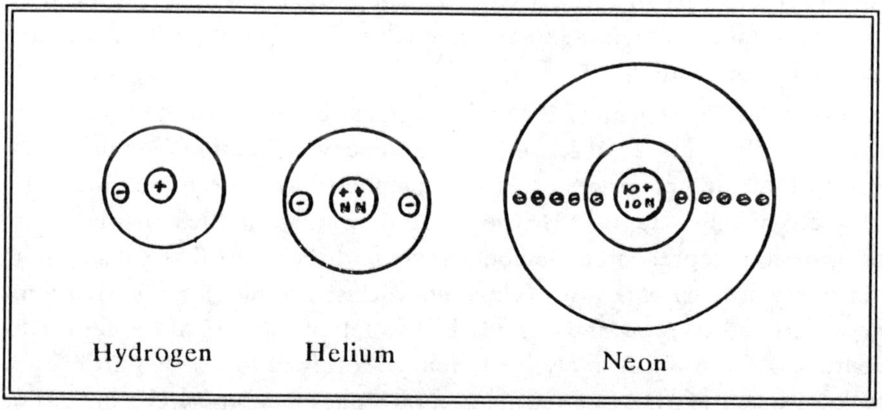

Fig. 9.2. Representations of models of atoms.

The mass of the proton has been determined as 1.67×10^{-24} grams, which is very close to the mass of the neutron. Electrons, on the other hand, have a mass of 9.1×10^{-28} grams, which is three orders of magnitude less than the mass of either the proton or neutron. Because neutrons and protons exist only in the nucleus, the nucleus constitutes the predominance of atomic mass and, for chemical reactions, the mass of the electrons are ignored. The number of protons and neutrons in the nucleus of an atom is expressed as its "mass number."

Actual, more precise expressions of atomic mass are based on the mass of carbon 12, which is assigned an atomic mass unit (*amu*) of one. The mass determination of an element is given in terms of the number of carbon 12 masses. Actually, the atomic mass of an element is the average mass of the commonly occurring isotopes of the element. An isotope of an element is an atom of the same element with the same atomic number but a different atomic mass. Isotopes of an element differ only in the number of neutrons in their nuclei. For example, hydrogen has three common isotopes: protium, deuterium and tritium. All three have one electron and one proton. However, deuterium has two neutrons in its nucleus and tritium has three. Protium has none. Deuterium constitutes less than 0.02% of the hydrogen in nature; tritium is radioactive and appears in nature in very limited quantities.

Scientists look for order in nature so that, given certain characteristics or conditions, they can make predictions. It is much easier to find a rule that fits all known cases and can predict new cases rather than construct volumes of data tables designed to incorporate all cases. Data tables do not predict.

Mankind has struggled through the ages to identify patterns among the basic elements. In 1869 Dmitri Mendeleev, a Russian chemist, suggested that the properties of various elements were periodic functions of their atomic masses. He postulated that anomalies in his mass progressions represented elements as yet undiscovered. The subsequent discovery of five *noble gases* (elements with stable outershells – helium, neon, argon, krypton and xenon) in the latter part of the nineteenth century and the X-ray analysis of Henry Mosely in the early part of the twentieth century led to modification of Mendeleev's periodic law. The natural characteristics of the elements turned out to be better explained as functions of their atomic numbers rather than atomic mass as suggested by Mendeleev.

The Periodic Table

Figure 9.3 is a modern table, known as the *periodic table*. In it the elements are arranged in order of their atomic number. The vertical columns represent groups or families of elements which have similar characteristics. The horizontal rows are referred to as *periods* or *series*. The row numbers indicate elements with electrons in the same energy levels. All the so-called noble gases in group 18 have their highest energy levels filled, thereby completing each respective period.

In general, the elements on the left-hand side of the table are metals and those on the right-hand side are non-metals. The elements along the heavy dark line in a staircase pattern (running from Boron [at no. 5] to Astatine [at no. 85]) are known as metalloids that show both metallic and non-metallic properties.

Groups are divided into *s, p, d* and *f* blocks. The letter designation is assigned to different shapes of the orbitals, some of which are three dimensional. Again, one is cautioned about a two-dimensional model depicting orbitals in a single plane.

Special groups have been identified on the basis of specific characteristics. The transition elements fill electrons in the *d*-shaped orbitals. The earth elements of the Lanthanoid series have very similar chemical properties and were at one time considered rare. The Actinoid series has characteristics similar to the period 6 elements, beginning with the element actinium.

Individual blocks for each element contain the chemical symbol, the atomic number and the atomic mass of that element. The electron energy level configuration, atomic radius and various oxidation states in which an atom of the element can exist are also included.

Bonding

The combining of elements involves only the outer energy levels of their atoms. In general, atoms in groups 16 and 17 have an affinity for electrons while the rest tend to give up electrons. Some, under certain conditions, will exhibit both properties.

Periodic Table of the Elements

Key:

Atomic Mass	Symbol	Selected Oxidation State	Atomic Number	Electron Configuration
12.0111	C	−4, +2, +4	6	[He] $2s^2 2p^2$

GROUPS

s-block (groups 1–2); d-block — Transition Elements (groups 3–9)

Period	1	2	3	4	5	6	7	8	9
1	**H** 1 1.00794 +1, −1 $1s^1$								
2	**Li** 3 6.941 +1 [He] $2s^1$	**Be** 4 9.01218 +2 [He] $2s^2$							
3	**Na** 11 22.98977 +1 [Ne] $3s^1$	**Mg** 12 24.305 +2 [Ne] $3s^2$							
4	**K** 19 39.0983 +1 [Ar] $4s^1$	**Ca** 20 40.08 +2 [Ar] $4s^2$	**Sc** 21 44.9559 +3 [Ar] $3d^1 4s^2$	**Ti** 22 47.88 +2, +3, +4 [Ar] $3d^2 4s^2$	**V** 23 50.9415 +2, +3, +4, +5 [Ar] $3d^3 4s^2$	**Cr** 24 51.996 +2, +3, +6 [Ar] $3d^5 4s^1$	**Mn** 25 54.9380 +2, +3, +4, +7 [Ar] $3d^5 4s^2$	**Fe** 26 55.847 +2, +3 [Ar] $3d^6 4s^2$	**Co** 27 58.9332 +2, +3 [Ar] $3d^7 4s^2$
5	**Rb** 37 85.4678 +1 [Kr] $5s^1$	**Sr** 38 87.62 +2 [Kr] $5s^2$	**Y** 39 88.9059 +3 [Kr] $4d^1 5s^2$	**Zr** 40 91.224 +4 [Kr] $4d^2 5s^2$	**Nb** 41 92.9064 +3, +5 [Kr] $4d^4 5s^1$	**Mo** 42 95.94 +3, +6 [Kr] $4d^5 5s^1$	**Tc** 43 (98) +4, +6, +7 [Kr] $4d^5 5s^1$	**Ru** 44 101.07 +3 [Kr] $4d^7 5s^1$	**Rh** 45 102.906 +3 [Kr] $4d^8 5s^1$
6	**Cs** 55 132.905 +1 [Xe] $6s^1$	**Ba** 56 137.33 +2 [Xe] $6s^2$	**La-Lu** 57 - 71	**Hf** 72 178.49 +4 [Xe] $4f^{14} 5d^2 6s^2$	**Ta** 73 180.948 +5 [Xe] $4f^{14} 5d^3 6s^2$	**W** 74 183.85 +6 [Xe] $4f^{14} 5d^4 6s^2$	**Re** 75 186.207 +4, +6, +7 [Xe] $4f^{14} 5d^5 6s^2$	**Os** 76 190.2 +3, +4 [Xe] $4f^{14} 5d^6 6s^2$	**Ir** 77 192.22 +3, +4 [Xe] $4f^{14} 5d^7 6s^2$
7	**Fr** 87 (223) +1 [Rn] $7s^1$	**Ra** 88 226.025 +2 [Rn] $7s^2$	**Ac-Lr** 89 - 103	**Unq*** 104 (261)	**Unp** 105 (262)	**Unh** 106 (263)	**Uns** 107 (262)	**Uno** 108	**Une** 109

*The systematic names and symbols for elements of atomic numbers greater will be used until the approval of trivial names by IUPAC.

d-block

La 57 138.906 +3 [Xe] $5d^1 6s^2$	
Ac 89 227.028 +3 [Rn] $6d^1 7s^2$	

Lanthanoid Series / Actinoid Series

Series					
Lanthanoid Series	**Ce** 58 140.12 +3, +4 [Xe] $4f^2 6s^2$	**Pr** 59 140.908 +3 [Xe] $4f^3 6s^2$	**Nd** 60 144.24 +3 [Xe] $4f^4 6s^2$	**Pm** 61 (145) +3 [Xe] $4f^5 6s^2$	**Sm** 62 150.36 +2, +3 [Xe] $4f^6 6s^2$
Actinoid Series	**Th** 90 232.038 +4 [Rn] $6d^2 7s^2$	**Pa** 91 231.036 +4, +5 [Rn] $5f^2 6d^1 7s^2$	**U** 92 238.029 +3, +4, +5, +6 [Rn] $5f^3 6d^1 7s^2$	**Np** 93 237.048 +3, +4, +5, +6 [Rn] $5f^4 6d^1 7s^2$	**Pu** 94 (244) +3, +4, +5, +6 [Rn] $5f^6 7s^2$

Conforms to the recommendations of the New York State Department of Education.

s-block

18
4.00260 0 **He** 2 $1s^2$

p-block

10	11	12	13	14	15	16	17	18
			10.81 +3 **B** 5 [He] $2s^22p^1$	12.0111 −4 +2 +4 **C** 6 [He] $2s^22p^2$	14.0067 −3 −2 −1 +1 +2 +3 +4 +5 **N** 7 [He] $2s^22p^3$	15.9994 −2 **O** 8 [He] $2s^22p^4$	18.998403 −1 **F** 9 [He] $2s^22p^5$	20.179 0 **Ne** 10 [He] $2s^22p^6$
			26.98154 +3 **Al** 13 [Ne] $3s^23p^1$	28.0855 −4 +2 +4 **Si** 14 [Ne] $3s^23p^2$	30.97376 −3 +3 +5 **P** 15 [Ne] $3s^2 3p^3$	32.06 −2 +4 +6 **S** 16 [Ne] $3s^23p^4$	35.453 −1 +1 +3 +5 +7 **Cl** 17 [Ne] $3s^23p^5$	39.948 0 **Ar** 18 [Ne] $3s^23p^6$
58.69 +2 +3 **Ni** 28 [Ar] $3d^84s^2$	63.546 +1 +2 **Cu** 29 [Ar] $3d^{10}4s^1$	65.39 +2 **Zn** 30 [Ar] $3d^{10}4s^2$	69.72 +3 **Ga** 31 [Ar] $3d^{10}4s^24p^1$	72.59 −2 +2 +4 **Ge** 32 [Ar] $3d^{10}4s^24p^2$	74.9216 −3 +3 +5 **As** 33 [Ar] $3d^{10}4s^24p^3$	78.96 −2 +4 +6 **Se** 34 [Ar] $3d^{10}4s^24p^4$	79.904 −1 +1 +5 **Br** 35 [Ar] $3d^{10}4s^24p^5$	83.80 0 +2 **Kr** 36 [Ar] $3d^{10}4s^24p^6$
106.42 +2 +4 **Pd** 46 [Kr] $4d^{10}5s^0$	107.868 +1 **Ag** 47 [Kr] $4d^{10}5s^1$	112.41 +2 **Cd** 48 [Kr] $4d^{10}5s^2$	114.82 +3 **In** 49 [Kr] $4d^{10}5s^25p^1$	118.71 +2 +4 **Sn** 50 [Kr] $4d^{10}5s^25p^2$	121.75 −3 +3 +5 **Sb** 51 [Kr] $4d^{10}5s^25p^3$	127.60 −2 +4 +6 **Te** 52 [Kr] $4d^{10}5s^25p^4$	126.905 −1 +1 +5 +7 **I** 53 [Kr] $4d^{10}5s^25p^5$	131.29 0 +2 +4 +6 **Xe** 54 [Kr] $4d^{10}5s^25p^6$
195.08 +2 +4 **Pt** 78 [Xe] $4f^{14}5d^96s^1$	196.967 +1 +3 **Au** 79 [Xe] $4f^{14}5d^{10}6s^1$	200.59 +1 +2 **Hg** 80 [Xe] $4f^{14}5d^{10}6s^2$	204.383 +1 +3 **Tl** 81 [Xe] $4f^{14}5d^{10}6s^26p^1$	207.2 +2 +4 **Pb** 82 [Xe] $4f^{14}5d^{10}6s^26p^2$	208.980 +3 +5 **Bi** 83 [Xe] $4f^{14}5d^{10}6s^26p^3$	(209) +2 +4 **Po** 84 [Xe] $4f^{14}5d^{10}6s^26p^4$	(210) **At** 85 [Xe] $4f^{14}5d^{10}6s^26p^5$	(222) 0 **Rn** 86 [Xe] $4f^{14}5d^{10}6s^26p^6$

than 103

Mass numbers in parentheses are those of the most stable or most common isotope.

f-block

151.96 +2 +3 **Eu** 63 [Xe] $4f^76s^2$	157.25 +3 **Gd** 64 [Xe] $4f^75d^16s^2$	158.925 +3 **Tb** 65 [Xe] $4f^96s^2$	162.50 +3 **Dy** 66 [Xe] $4f^{10}6s^2$	164.930 +3 **Ho** 67 [Xe] $4f^{11}6s^2$	167.26 +3 **Er** 68 [Xe] $4f^{12}6s^2$	168.934 +3 **Tm** 69 [Xe] $4f^{13}6s^2$	173.04 +2 +3 **Yb** 70 [Xe] $4f^{14}6s^2$	174.967 +3 **Lu** 71 [Xe] $4f^{14}5d^16s^2$
(243) +3 +4 +5 +6 **Am** 95 [Rn] $5f^77s^2$	(247) +3 **Cm** 96 [Rn] $5f^76d^17s^2$	(247) +3 +4 **Bk** 97 [Rn] $5f^97s^2$	(251) +3 **Cf** 98 [Rn] $5f^{10}7s^2$	(252) **Es** 99 [Rn] $5f^{11}7s^2$	(257) **Fm** 100 [Rn] $5f^{12}7s^2$	(258) **Md** 101 [Rn] $5f^{13}7s^2$	(259) **No** 102 [Rn] $5f^{14}7s^2$	(260) **Lr** 103 [Rn] $5f^{14}6d^17s^2$

Atoms that gain or lose electrons resulting in unbalanced charges are known as *ions*. Metals tend to have few electrons in their outer energy levels and tend to give off electrons, thus becoming *positively ionized*. Non-metals tend to have outer electron shells that are almost filled and have a tendency to take on electrons to fill these shells to become *negatively ionized.*

The process of bonding to form a compound can be illustrated by the bonding of sodium and chlorine to form sodium chloride (NaCl). Sodium has one electron in its outer level. When chlorine, with seven electrons in its outer shell, comes into the proximity of sodium and the proper energy is provided, the sodium electron is transferred to the chlorine atom and a negatively charged *chloride* ion is formed:

$$Na - 1\,e^{-} \longrightarrow Cl^{-}$$

Because the sodium atom lost one electron and now has one less negative charge than positive charges, it now exists as a positive sodium ion. The compound formed is $Na^{+}Cl^{-}$, *sodium choride:*

$$Na + Cl \longrightarrow Na^{+}Cl^{-}$$

The sodium and chlorine ions are now bonded through a transfer of electrons. Because ions were formed in the bonding process, it is known as *ionic bonding*. Ionic bonds result in a crystalline structure rather than a true molecular form. Individual ions cannot be directly associated with another specific ion, such as that found in molecular form.

Reacting gasses such as oxygen, hydrogen, nitrogen, chlorine and fluorine form bonds between pairs of atoms. Consider the hydrogen atom with one electron in its unfilled energy shell:

A more stable condition is achieved by sharing an electron with a

$$H + H \longrightarrow H{:}H$$

second hydrogen atom thereby forming a diatomic molecule with a formula:

$$H_2$$

Some common diatomic gases and their chemical formulas are listed in Table 9.1.

gas	formula
oxygen	O_2
hydrogen	H_2
nitrogen	N_2
flourine	F_2
chlorine	Cl_2
bromine	Br_2

Table 9.1. Common diatomic gases and their chemical formulas.

Most of the common elemental gases, outside of the noble gases, exist in the diatomic state. Such chemical bonding through electron sharing is known as covalent bonding.

Covalent bonding can take place between atoms of different elements. Nitrogen and hydrogen are united through a covalent bond in ammonia gas, as are carbon and hydrogen in methane gas. In fact, thousands of organized molecules of covalent hydrogen-carbon bonds have been synthesized in the plastics, petroleum and other petrochemical industries.

*****

Experiment 9–1. The Properties of Some Common Gases

You can gain a knowledge of some common gases by running simple tests on them. Four gases that can be readily obtained will be used in your investigations. The generating apparatus shown in Figure 9.4a will be used to generate gases *a, b* and *c.* Gas *d* will be obtained from the gas supply at your lab table.

Collect four test tubes of each gas and bubble some of the gas through limewater (Fig. 9.4b) to see if it produces a chalky-white color. Run the various tests on the gases collected and set up the apparatus for generation and tests on each successive gas.

Figure 9.5 provides a convenient way to systematize your observations. A gas may support combustion but not burn. Test this characteristic by lowering a glowing splint into the test tube and observe if it continues to glow. Flammability may be tested by lowering a burning splint into the gas.

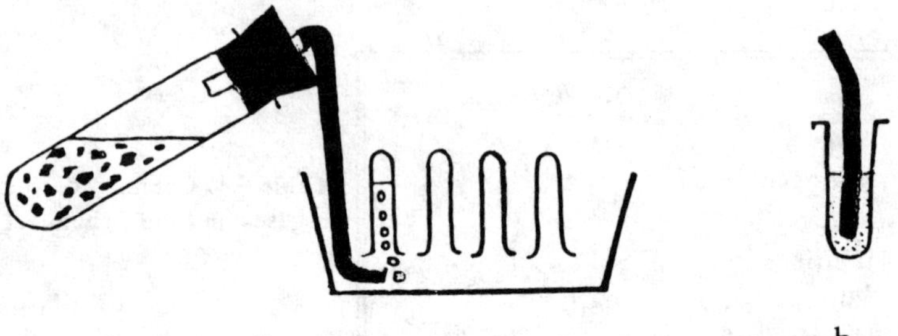

Fig. 9.4. Gas generation apparatus.

You can determine the relative density by placing a test tube of the gas directly over an open test tube of air for about 5 to 8 seconds. Place either a glowing or burning splint, whichever is appropriate, into the mouth of the test tube for indication of the gas (Fig. 9.6a). Presence of the gas with a density greater than air is indicated by a downward movement. In Figure 9.6b no appreciable rise in the gas confirms the observation of Figure 9.6a.

You can generate gas *a* by filling the generators one-third full of water.

gas	supports combustion	flammable	density vs. air		milky
			less	more	
a					
b					
c					
d					

Fig. 9.5. Test scheme for four gases.

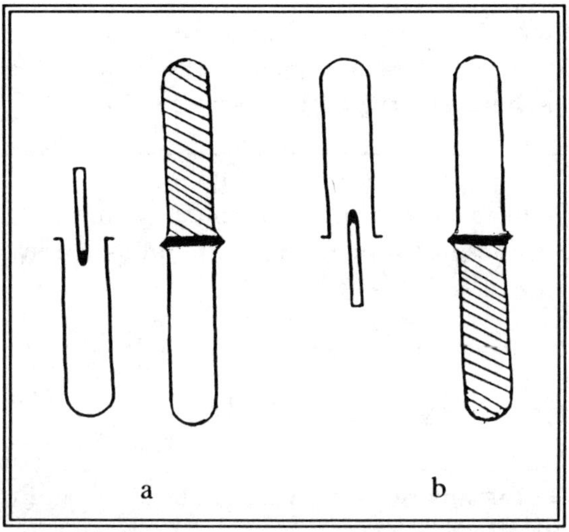

Fig. 9.6. Tests for relative density.

Drop in two halves of an Alka-Seltzer tablet, quickly restopper. Collect four test tubes of the gas and conduct the limewater test. Run the tests on the gas, clean out all of the apparatus, and prepare to generate gas *b*.

Gas *b* is generated by filling the generator half full of hydrogen peroxide. Gently warm the gas-generating tube with your hand (Caution) and collect the gas. Run the tests, clean out all of the apparatus and prepare to collect gas *c*.

Generate gas *c* by placing pieces of magnesium metal into the generator filled one-third full of dilute sulfuric acid.

CAUTION: Goggles must be worn. If any acid gets on your skin or in your eyes, flush with clean, cool, running water for several minutes and seek first aid.

Run the tests, clean out all of the apparatus and prepare to collect gas *d*. You will not need the generating test tube for gas *d*.

Gas *d* may be collected and the limewater test run by connecting the rubber delivery tube directly to a laboratory gas jet.

CAUTION: Considerable care should be exercised to insure the gas flows slowly, so unnecessary quantities of gas are not released into the atmosphere causing an explosive condition.

Compare the results of your tests with your classmates' results. Can you identify the gases? Keep these characteristics in mind as you will encounter them again.

*****

Combining Ratios of Elements

Mixtures can be concocted in any ratio desired. A certain brand of dog food may contain a specific percentage of protein, fat and carbohydrates, where a second brand may have a significantly different composition. Baking cookies with two pinches of salt rather than one pinch may not alter the cookies' composition such that you could taste the difference! In Experiment 9–2, you will gather some evidence as to whether the same is true for atomic combinations forming compounds.

*****

Experiment 9–2. The Decomposition of Water

Water cannot be broken down by the processes you have used to examine matter. In this experiment you will set up an apparatus which runs an electric current through water to break the water into its constituent parts.

Fill two test tubes with tap water, invert them in a beaker one-half full of water and arrange them as shown in Figure 9.7. Be sure there are no air bubbles in the test tubes and that the electrodes do not make contact with each other. Notice that the path for the charges from the battery can only pass from one battery terminal to one electrode, through the water to the opposite electrode and on to the opposite terminal of the battery. Any gas formed from the resultant transfer of energy will be collected in the test tubes above the electrodes in the water. It is important that you understand that the electric current flows

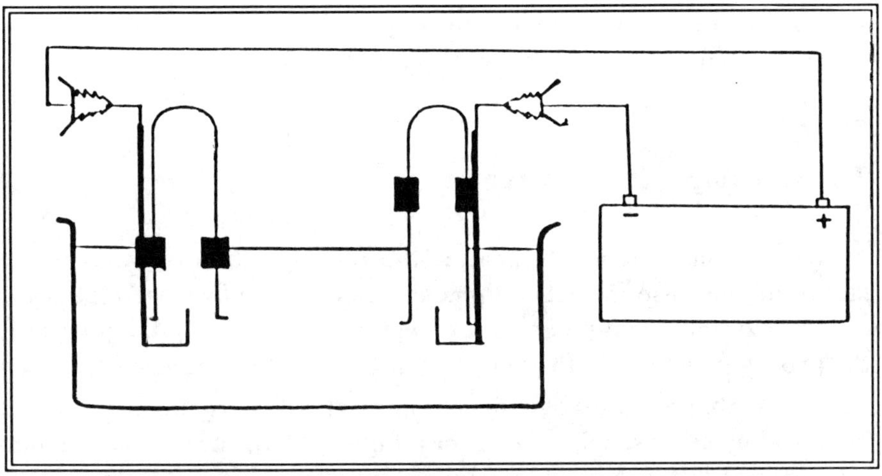

Fig. 9.7. Apparatus for Experiment 9–2.

through the portion of the water between the electrodes. Connect the wire leads to the terminals of the battery and observe any activity at the electrodes. Add a few teaspoons of sodium carbonate to the water and stir with a plastic spoon to hasten the dissolving.

CAUTION: Use only a plastic spoon so you will not short out the electrodes.

Observe any change in activity at the electrodes. Next, add another battery to the circuit and observe the activity at the electrodes. Did the addition of sodium carbonate make a difference? What about increasing the size of the battery?

Allow the current to run until *one* of the test tubes is completely filled with gas, at which time you should disconnect the leads from the battery. Carefully stopper the gases in the test tubes with the mouths of the tubes under the water. Note which electrode was connected to the positive and which to the negative terminal of the battery.

What can you say about the relative volumes of the two gases? What charges (+ or –) must have been on the atoms of each gas when they formed the water molecule? Justify your answer. Test the two gases first with a glowing splint (quickly restopper the gases) and then with a burning splint. Can you identify the gases?

*****

The Recomposition of Water

You will not attempt to analyze the recomposition of water in the laboratory because special equipment is required. If hydrogen is burned in pure oxygen, a minute amount of water vapor is formed. Apparently the process you used in the decomposition of water is reversible.

The combining ratios of hydrogen and oxygen in producing water can be tested by using special glass tubes. Figure 9.8 illustrates the combining ratios of hydrogen and oxygen in special glass tubes. The tubes are first filled with mercury and twenty milliliters of oxygen are inserted into each tube (Fig. 9.8a). Figure 9.8b represents the volumes of gases after 20 ml, 40 ml and 60 ml of hydrogen gas have been respectively added to tubes 1, 2 and 3.

Figure 9.8c represents the remaining volumes of gases after ignition. Notice there is no longer any gas appearing in tube #2. Testing the remaining gases indicates 10 ml of oxygen in tube #1 and 20 ml of hydrogen remaining in tube #3. Table 9.2 shows the combining volumes of the gases to form water. The amount of water produced is very minute. The ratio of hydrogen to oxygen in tube 1 is 10:10 = 1:1; tube 2 is 40:20 = 2:1; tube 3 is 30:10 = 3:1. The combining ratio by volume of hydrogen to oxygen always comes out to be 2:1 in the formation of water, irrespective of what the ratios of the gas mixtures were to begin with. A similar test could be run by using larger amounts of oxygen with fixed amounts of hydrogen, but it wouldn't seem logical because the combining amount of hydrogen is always greater than the amount of oxygen. Experiments conducted by different investigators with a variety of materials led to the conclusion that elements combine in a constant proportion that can be expressed as ratios of small whole numbers. In fact, a chemical compound can be defined as two or more elements chemically united in definite proportions.

*****

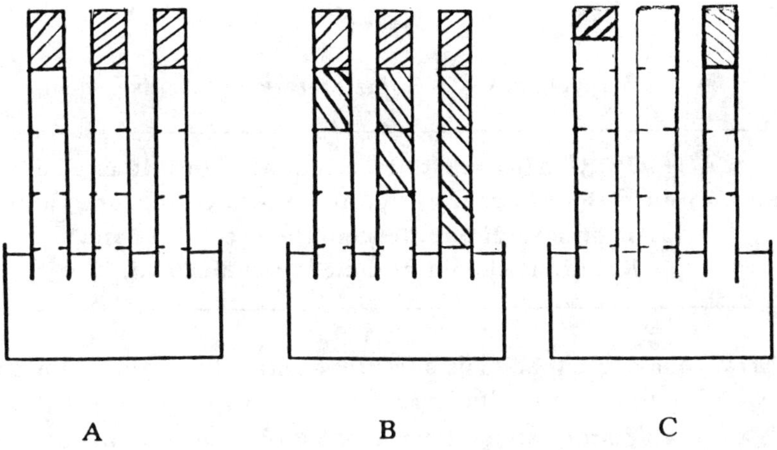

Fig. 9.8. Combinations of hydrogen and oxygen forming water (a) 20 ml oxygen, (b) 20 ml, 40 ml and 60 ml H_2 respectively added to each tube and (c) after ignition.

Two elements may combine under different conditions to form different compounds. For example, hydrogen and oxygen do not always form water. Under special conditions they may be made to form hydrogen peroxide. However, the ratio of hydrogen to oxygen in

	Tubes		
	1 (ml)	2 (ml)	3 (ml)
Oxygen	10	20	20
Hydrogen	20	40	40

Table 9.2. Combining volumes of the gases.

hydrogen peroxide will always be in a 2:2 ratio and when conditions are set to produce water, the ratio of hydrogen to oxygen in the water molecule will always be in a 2:1 ratio. The law of constant combining proportions is a powerful principle in science.

*****

Experiment 9–3. Some Metals and Acids

CAUTION: Goggles must be worn at all times. If an acid gets on your skin or in your eyes, flush it with clean, cool, running water for several minutes and then seek first aid. Reactions should be done in a fume hood.

Arrange in a test-tube rack, a test tube with dilute hydrochloric acid, one with dilute (0.1-N) sulfuric acid and one with dilute (0.1-N) nitric acid, all to a depth of about 2 cm. Place a piece of zinc in each of the acids and record your observations.

Flush out the test tubes and repeat the process with magnesium metal. Again flush the test tubes and repeat the process with copper metal. Warm the test tube this time by holding it in your hand to add energy to the system. Did the acids all behave in the same way? Were the products the same? Could you recover the metals? Try it with the solution which resulted from the zinc/hydrochloric acid reaction solution by gently warming a droplet on a microscope slide until dry. Is there any evidence of a solid in the solution? What were the results?

*****

Experiment 9–4. Copper Oxide Reduction

Charcoal has been used through the ages to recover pure metals from their ores. The iron and steel industry uses tons of coke to serve the purpose of charcoal. In this experiment you will use charcoal mixed with an oxide of copper to try to recover the copper.

Carefully measure about 3 g of copper oxide and thoroughly mix it with about 0.2 g of finely pulverized charcoal. Determine the mass of the

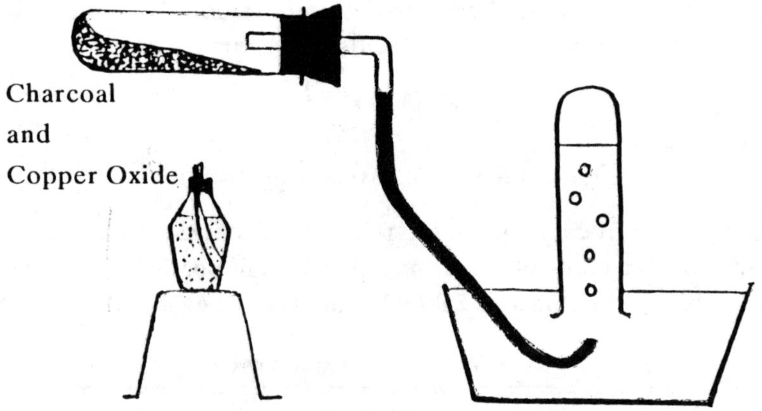

Fig. 9.9. Apparatus for Experiment 9–4.

copper, charcoal and the test tube. Place the mixture in a test tube set up in the tube arrangement shown in Figure 9.9.

Heat the mixture gently at first and then vigorously until the mixture glows brightly. Collect the gas and test it. Examine the copper oxide-charcoal mixture. Do you see any evidence of copper?

Determine the mass of the remaining mixture and the test tube. Was the mass reduced or increased? Explain why. Can you identify the gas given off?

CAUTION: Goggles must be worn. Do not aim the mouth of a test tube toward yourself or anyone else in the room.

*****

Acids and Bases

People tend to label liquids that dissolve hair or animal fibers or attack the skin as acids. Most of these are not acids. Certain materials used to clean drains also behave similarly when exposed to organic material. In Experiment 9–5 you will investigate different materials called *hydroxides* as well as *acids* to determine similarities and differences, using litmus paper to identify acids and hydroxides.

*****

Experiment 9-5. Interesting Juices

Place a drop of each one of the following materials on a microscope slide and touch a piece of litmus paper to a small portion of the droplet. Make a table of the items in Table 9.3 and record each color change.

Table 9.3. Materials for Experiment 9–5.

Materials	
1. orange juice	9. lime juice
2. lemon juice	10. Draino
3. Coca Cola	11. hair shampoo (A)
4. limewater	12. hair shampoo (B)
5. milk of magnesia	13. Alka-seltzer water
6. vinegar	14. hydrochloric acid
7. ammonia water	15. sulfuric acid
8. tap water	16. saliva

CAUTION: Goggles are required.

You should find that some of the materials change the color of the litmus paper to blue and some change it to red. How would you label all of the substances in one group? What would you call the substances in the other group?

Place 50 cm^3 of distilled water in a clean, small beaker and add four drops of bromothymol blue (BTB).

Place a clean piece of glass tubing in the beaker and gently bubble your breath into the water via the glass tube. Do you eventually notice any color change in the water? What causes it? Test the water with a piece of blue litmus paper. Do the results give you a clue? What is one of the gases you exhale? What relationship does it have to the color change?

*****

Experiment 9–6. Draino and an Acid

CAUTION: Goggles required. If you get any hydrochloric acid solution or Draino solution on your skin or in your eyes, flush it with cool, running water for several minutes and seek first aid. Continue flushing any Draino solution on the skin until the slippery feeling disappears.

Use a medicine dropper to place exactly 10 drops of 0.1-N hydrochloric acid and 10 drops of 0.1-N Draino solution in a clean, dry evaporating dish. Add about 5 cm^3 of distilled water. Flush the stirring rod with distilled water and thoroughly stir the mixture. Test the mixture with litmus paper. Is it an acid or something else?

Pour off half of the liquid and evaporate two drops of the remaining liquid to dryness by gently warming on a microscope slide. What are some of characteristics of the solid? The solid is sodium chloride, NaCl (common table salt).

CAUTION: Do not taste any materials in the laboratory!

Acids are neutralized by materials that turn litmus paper blue. Such substances are labeled *bases, hydroxides* or *alkalis*. Acids and bases neutralize each other to form a salt (metal and non-metal) and water.

Acids form hydronium ions (H_3O^+) in water, whereas hydroxides form OH^- ions. Molecules are said to *dissociate* when ionized in water. The term *hydroxide* is usually incorporated into the name of a base, e.g., sodium hydroxide.

A numerical method for expressing the acidity or alkalinity of a solution that is based on the hydronium ion concentration of a solution is known as the *pH* of a solution. More specifically, the pH of a solution is the reciprocal of the exponent of the hydronium ion concentration expressed in powers of ten. One way the pH of a solution can be determined is through the use of hydrion paper, which contains a dye that responds to a specific pH with a specific color change. Use some hydrion paper to check some of the shampoos used in Experiment 9–3. What does the pH of a shampoo contribute to its effectiveness?

*****

This chapter began with a discussion of the basic nature of matter as developed through the ages and resulting in our present theories of matter. Periodicity and the basic structure of elemental substances were expanded upon.

Two basic forms of bonding hold the atoms of elements together. *Ionic bonds*, which involve the transfer of electrons, result in compounds with a crystalline structure. *Covalent bonds* are based on the sharing of electrons and form diatomic molecules and compounds of constituent molecules. The covalent bonding of carbon and hydrogen is the foundation for the gigantic petrochemical industries.

Basic to the science of interacting atoms is the *law of definite proportions.* Atoms unite to form molecules only in ratios of small whole numbers. Compounds may be broken down into constituent parts, in the same proportions as when the elements originally combined. If amounts of the combining elements are provided in ratios that violate the law of definite proportions during the formation of compounds, only the amounts of each element that obey the law of definite combining proportions will combine. Nature cannot be persuaded to violate this law.

Radioactivity

During the latter part of the nineteenth century, the element uranium, found in a brownish-black mineral called *pitchblende*, was used for coloring porcelain and glass. Henri Becquerel, a French scientist, experimented with compounds of uranium in 1896. He was interested in these compounds because they glowed in the dark. It was Becquerel's belief that the glow was activated by ultraviolet light emitted from sunlight. He observed that placing these compounds on a photographic plate wrapped in dark paper, thick enough not to be affected by bright sunlight, revealed images of the compounds when the plates were developed. Becquerel believed the images were due to radiation, similar to X-rays, which were somehow triggered by sunlight.

To eliminate the possibility of the photographic plates being affected by a gas given off from the substances due to the heat of sunlight, Becquerel placed a glass plate between the samples and the plates, with no effect. He conducted other experiments, substituting metal plates for the glass plates. Following his belief that the radiation of the compounds originated from sunlight, he experimented with reflected sunlight as well as light refracted through a glass prism.

During Becquerel's experiments with uranium and sunlight, he ran into several days of cloudy weather. He placed the wrapped photographic plates with the uranium compounds attached in a cabinet until he could resume his activities. Days later, he developed the photographic plates. To his amazement, the plates showed the images of the compounds. Becquerel realized there was no connection between the radiation and sunlight. In fact, the radiation was spontaneous and was not affected by light, heat or pressure.

Later, in 1899, Earnest Rutherford (1871–1937), a British physicist, identified two forms of radiation given off from uranium. One type of radiation that was stopped by a thin sheet of paper or aluminum, he called *alpha* radiation. The second type, which had about one hundred times the penetrating power of alpha, he called *beta* radiation. A third type of radiation, similar to X-rays and more penetrating than beta rays, was discovered in 1900 by French chemist Paul Villard and named *gamma rays*.

Upon the suggestion of Becquerel, Pierre and Marie Curie investigated the naturally-occurring radiation (*radioactivity*) of pitchblende ore. They observed that its radiation was stronger than uranium and correctly believed that it contained one or more radioactive substances besides uranium.

Using a series of tests, many of which were similar to ones you have used to identify characteristic properties, the Curies were able to isolate a quantity of material so small that it was detectable only by its radioactivity—not by an optical or chemical analysis. Chemically, it behaved as a compound of bismuth. Eventually, a sample was isolated and yielded to spectroscopic analysis. Its spectral lines were unique and it was named *polonium* after Poland, the nation of Marie Curie's birth.

A second fraction of the pitchblende samples, which contained no polonium, indicated strong radioactivity. Further analysis in 1898 revealed a second radioactive element, radium, which was much more radioactive than uranium. Rutherford discovered that the radioactive elements thorium and radium gave off a radioactive gas. This "radium gas" was difficult to analyze because it was chemically inert. The British chemists William Ramsay (1852–1916) and Frederick Soddy (1877–1956) were able to liquefy the gas and eventually identified it as a chemical element. It became known as *radon*.

Radon gas was unique, not only because of its radioactivity but also because, as its spectral lines steadily grew weaker, the spectral lines of another element began to appear and gain in intensity. The new spectral lines were identified as those of the element helium. Soddy and Ramsay concluded that the element radon spontaneously changed into the element helium. It was eventually shown that the alpha particles observed by Rutherford were actually the nuclei of helium atoms and the beta rays were electrons ejected from a radioactive nucleus. Helium was found to be associated with the radioactive decay of polonium, radium, radon, thorium and uranium.

Researchers also observed that a small amount of lead was associated with radioactive materials and later discovered that all naturally-occurring substances eventually form stable, non-radioactive lead. Radioactive disintegration is the only example of one element changing into another element.

The Nature of Radioactivity

Radioactive materials have unique characteristics. They spontaneously and continuously emit particles from their unstable nuclei, giving off energy and yielding simpler elements. This nuclear decay is apparently due to certain ratios of protons and neutrons (as well as a variety of other particles) at various energy levels within the nucleus causing the element to be unstable.

The disintegration of the nucleus is not only spontaneous but is probabilistic. A sample of radioactive substance consists of many atoms, each disintegrating at random. One reading from a detector near the sample will indicate a very erratic number of particles being emitted with no apparent pattern. However, if a large number of "counts" are made over a relatively long period of time, a pattern emerges, such that predictions of particle emanations can be made. An analogy is the tossing of a coin. A relatively small number of tosses may show more "heads" than "tails." However, a very large number of tosses will demonstrate a fifty-fifty chance of the coin landing either "heads" or "tails."

Because nuclear disintegrations are probabilistic, they are not influenced by outside factors. They are an intrinsic property dictated by the construct of the atom. You cannot stop it, speed it up or slow it down. The combining of radioactive elements to form compounds does not alter the radioactive characteristics of an element, even though their chemical characteristics are altered.

Energy also accompanies nuclear disintegration. A tube of radon gas, for example, will always maintain a temperature slightly higher than room temperature. Energy is given off as it emits particles in changing to helium. It is possible to harness this energy in so-called nuclear reactors. Uncontrolled critical amounts of nuclear energy can cause mass destruction, as in the case of nuclear bombs or the failure of nuclear power plants.

Due to the emission of particles and gamma rays, radioactive materials affect photographic emulsions, cause some substances (such as certain sulfides) to glow and ionize gases. The latter characteristic led to the development of the Geiger counter, which consists of a glass tube containing a low-pressure gas separating two electrodes. Penetration of a disintegration particle ionizes the gas, causing a small current flow. The sound created is amplified and heard as a "click" in headphones and is registered on a meter. Several other indicators, such as scintillation counters, electroscopes and cloud chambers, utilize the ionization characteristic of radioactive materials.

Radioactive materials also cause physiological effects in living organisms. This characteristic can be beneficial as well as detrimental. Too much radiation can, for example, cause cancer, but controlled doses of radiation can destroy cancer cells. Radioactive "tracers" or chemicals that have been made radioactive artificially can be injected into the human bloodstream to help diagnose human diseases or malfunction.

All elements with an atomic number greater than 83 are radioactive. The elements polonium, radon, radium, actinium, thorium, protactinium and uranium are naturally radioactive. The rest have been produced by nuclear bombardment.

Half-life

Each radioactive element has a characteristic *half-life*, which is the time it takes for one-half of its mass to disintegrate. Half-lives of elements can range from days to thousands of years. For example, the half-life of an isotope of polonium is slightly over 138 days, for bismuth it is over 1018 years, and for radium, 1620 years. In other words, only one-half of the base of a given sample of radium would remain after 1620 years, one-fourth of the sample would remain after 3,240 years, and after 4,860 years, one-eightieth of the radioactive sample would remain. Some of the materials made radioactive artificially have half-lives of only a few seconds.

The concept of half-life is enlisted in the determination of the age of old materials. All living organisms and the remains of formerly living organisms contain the element carbon. Based on the assumption that the radioactive disintegration of elements remains the same across time, the amount of radioactive carbon remaining in a material is compared to its natural half-life, and its age can be computed by extrapolating back.

Artificial Radioactivity

Non-radioactive elements can be made radioactive by "firing" particles into their nuclei. Thus, they are rendered unstable and begin to disintegrate. At first, alpha particles were used because they were easy to obtain and accelerate. Rutherford used the alpha particles of radium to bombard nitrogen atoms to produce the first artificial radioactivity. Because alpha particles possess an electric charge, they require high levels of energy to penetrate an atomic nuclei. In 1931 Ernest O. Lawrence (1901–1958) developed the cyclotron. This mechanism is comprised of an oscillating electric field in a strong permanent magnetic field. Alpha particles introduced into the cyclotron are given pulses of energy at precise moments as they spiral through the magnetic field. This allows them to accelerate with sufficient energy to penetrate the nuclei of atoms in a target material. The betatron, synchrotron and linear accelerator are other devices developed to accelerate particles for nuclear bombardment for the study of atomic nuclei. A tremendous amount of information about the atom has been accumulated by studying the paths of particles leaving nuclei struck by these "atomic bullets." Some particle accelerators, such as the recent, federally-funded "super-conducting-super collider," are huge, occupying many acres of land.

Nuclear Fission and Fusion

When certain nuclei, such as the isotope uranium 235 (mass units), are bombarded with neutrons of appropriate energy, they become very unstable, eventually splitting. Each nuclear split generates free neutrons, which in turn may be absorbed by other uranium atoms. If a "critical mass" of uranium is involved, a sustained "chain reaction" results. If left uncontrolled, a nuclear explosion occurs. Employing movable rods of neutron-absorbing materials, such as boron or graphite, allows for a "controlled reaction" as in a nuclear reactor. It should be noted that the energy of a nuclear reactor cannot be tapped directly to produce electricity. Instead, the heat of the reaction is used to create steam that powers turbines coupled to electrical generators.

Fusion is somewhat the opposite of fission. In fusion, nuclei are combined to form heavier nuclei; whereas fission involves the disintegration of nuclei. Fusion does not require a critical mass because it is not a chain reaction dependent on a critical number of neutrons. Tremendous amounts of energy are required to initiate fusion reactions, and special conditions are necessary for them to be maintained. The sun derives its energy from the fusion of hydrogen into helium.

Sustained, controlled, fusion reactions have been attained for brief moments in the laboratory but have not been developed on a practical basis. Nuclear fusion holds considerable promise because the fuel can theoretically be deuterium oxide, *heavy water*, which is available from ordinary water. As well, a fusion reaction is "clean" and does not produce the radioactive waste products of a fission reaction.

Science/Technology/Society

For centuries human beings have contemplated the building blocks of nature. Aristotle influenced intellectual thought for centuries with the concept of the four basic substances of fire, water, earth and air. The alchemists were involved in a pseudochemistry based on folklore and mysticism. Modern science evolved into an empirically-based process that has now identified over one hundred chemical elements. Many benefits to humans have been derived from the analysis and synthesis of chemical compounds, particularly organic compounds. Other industries that have benefited include mining, metallurgy and petroleum refining.

Most, if not all, of the various chemical industrial processes involve by-products which are detrimental to the quality of life most people are aspiring to enjoy. By the same token, society is constantly plagued with the problem of finding sufficient supplies of natural resources and energy to fuel industry. The waste products of industry enter either the atmosphere or water resources. Eventually, they end up in surface water and/or ground water.

Processes that allow ores of metals to be converted to pure metals, usually in the form of oxides, lead to by-products such as sulfur and carbon dioxide that are expelled into the atmosphere. These gases combine with rain water much the same way that you formed carbonic acid by exhaling carbon dioxide into water. Carbonic acid and sulfuric acid are components of "acid" rain. This rain runs into streams and lakes and percolates into the ground water, the source for wells. These materials tend to precipitate out of the atmosphere, and with certain atmospheric conditions, they are trapped into temperature inversions, causing "smog."

Digging ores from the earth leaves ugly scars on its surface and creates erosion problems. Soil erosion also contributes to water pollution. Areas of the Great Lakes are being polluted by taconite tailings being dumped onto the shores where leeching carries impurities into the lakes and ground water.

Plastics can be a mixed blessing. On the one hand, they replace the use of metals in consumer goods, thereby conserving valuable, irreplaceable resources. The pollution resulting from the processing of ores is also eliminated through this substitution. The negative side of producing plastics is the pollution of the environment by the by-products of the plastics industry. Plastics are used to make many trivial plastic toys, gadgets, and containers that are "throw-away" items. As an example, consider the disposable cigarette lighter, flashlight and plastic container. Such items, which are not biodegradable and take years to decompose, pile up in landfills and are an eyesore along streets, roadways and in streams and lakes. Plastics also place a drain on critical petroleum resources and require energy to produce.

Metals have the advantage of being relatively easy to recycle into pure metal. States that have placed a monetary charge with redemption privileges on aluminum cans and plastic bottles have virtually eliminated that form of pollution, have generated a renewable resource and have provided opportunities for service and income to youth groups who collect and sell them.

The metals in discarded automobiles and other forms of scrap metal, including metallic wastes from manufacturing, can be recycled. Again, recycling causes energy drains and pollution side effects, but in many cases, these are less severe than beginning with original ores.

All citizens should be concerned with the problems of pollution, but some of the solutions result in inconveniences for individuals. If enough individuals become displeased with a law designed to alleviate pollution, it is possible they will bring pressure against government agencies to have the laws repealed. For example, recycling glass, metal and plastic containers requires cleaning and storage until they can be returned to a store or recycling center. It can also be messy, because the residual sugars and malts from the beverages are smelly and attract insects and vermin. Citizens need to focus on the overall impact on the environment and the quality of life. Some inconvenience is bound to be coupled with efforts to keep the environment from becoming polluted to the point life becomes unbearable. Citizens need to be well informed of the overall benefits derived from measures responsible for their inconvenience.

An example of the public making political choices on the basis of convenience and facility is the alteration of the 55-miles-per-hour speed restriction on the interstate highway system, which originally was put in place because of the 1973 oil embargo. The 55-miles-per-hour speed limit conserved energy and saved many human lives. It is well-established that high-speed driving requires quicker reflexes than low-speed driving and that a higher percentage of people are injured or killed as the speed limit increases. Recently, however, the speed limit was increased to sixty-five miles per hour on the "rural interstate highways." As a result, certain insurance companies raised their rates in recognition of increased property damage and personal injury. A trip on a "rural interstate" reveals automobiles traveling at 70 and 75 miles per hour. The increased speed causes higher consumption of fuel, with an associated increase in pollution and higher personal and property damage. Did government representatives make a wise and informed choice or a decision based on the influence of certain pressure groups? Some media presentations have suggested that the loss in human lives and increased cost of fuel is far offset by the increase in productivity of industry in moving goods across the highways faster.

The chemical industry has become an integral part of modern society in facilitating the production of the large amounts of food required to sustain large populations. Pesticides, herbicides and fertilizers are necessary for the production of concentrated plant growth. Growth hormones and pharmaceuticals are now used to sustain animal growth and prevent disease epidemics in confinement feedlots. The industrial wastes resulting from the production of these materials leads to disposal and energy utilization problems. The use of fertilizers, pesticides and herbicides placed on the soil also leads to potential lake and stream pollution and invasion of ground water, as does the erosion of animal waste products from animal feedlots.

Nuclear wastes present special hazards in that many of them are radioactive and have long half-lives. They need to be stored in special containers and stored in locations such that, if their containers do leak, they will not percolate into the ground water or end up in surface bodies of water.

Nuclear reactors are seen by many as the answer to the modern industrial society's insatiable appetite for energy. Others are extremely sensitive to the potential hazards of a nuclear fission plant that is not designed or operated properly. Nuclear fission catastrophes, such as the Three Mile Island power plant in the United States and the disaster of the Chernobyl plant in the Soviet Union, are offered as examples of resultant human suffering and what could happen. The question is: Has the paranoia concerning nuclear fission power plants developed to the point that it is virtually impossible to place new power plants on line, thus continuing a dependence on high-pollution, fossil-fuel plants?

Nuclear fusion power plants hold great promise for cheap, clean energy. Research is being conducted to bring about their reality. Is our society placing enough emphasis on their development? Could nuclear fusion bring about energy independence? What are the by-products of nuclear fusion?

You can see from the previous examples and discussion that the chemical industry, as other science and technology-driven industries, is a mixed blessing. Modern society cannot function without it, but by the same token, chemical waste disposal is an ever-increasing problem. It is important that the population be aware of the problems in order to deal with these problems rationally and in a manner that leads to optimum benefits for society, without inhibiting adequate industrial production.

Problems

1. According to what property of the various elements does the periodic table arrange specific elements?
2. How many electrons are contained in an orbital of an atom?
3. What is the maximum number of electrons which can be present in the first four energy levels of an atom?
4. Of what does most of an atom consist?
5. An atom of atomic number 5 has an atomic mass of 11. (a) How many protons does it have in its nucleus? (b) How many electrons does it possess? (c) How many neutrons does it possess, and (d) What is its net electrical charge?
6. The atomic mass of an atom is 108 and it has 61 neutrons in its nucleus. (a) What is its atomic number? (b) How many protons does it have in its nucleus? (c) What is its net electical charge? (d) How many electrons does it possess?

7. Describe tests which will tell if a gas (a) is flammable; (b) supports combustion; (c) is more or less dense than air; and (d) tell if the gas is carbon dioxide.
8. Name three characteristics of acids.
9. Name three characteristics of bases.
10. The pH of a certain material is 11. What can you say about the acidity of the material?

Chapter 10

Wave Motion and Sound

Waves

An old pun asks the question: What did the ocean say to the airplane? The equally asinine answer: It didn't say anything. It just waved on!

You are probably most familiar with the wave motions of water, but such terms as *waves of grain, flag waving, waves of soldiers* and *waves of insects* are common in the American language. Although some references to waves are real and some are metaphoric, all describe a disturbance in a medium (some form of matter).

Not all waves travel in a medium. For example, electromagnetic waves, such as those generated by a radio transmitter, travel through a vacuum and make it possible to communicate in outer space. In this chapter, however, you will deal mainly with waves in a medium to develop a wave model.

All waves are distortions of energy. As such, they consist of areas of high energy and areas of relatively low energy radiating from a specific point. This phenomenon can be observed by dropping a pebble into a calm body of water to generate the concentric waves represented in Figure 10.1. The disturbance moves out equally in all directions from *c*, the point of disturbance by the pebble. The waves appear as crests (high energy levels) and the corresponding troughs (low energy levels). The waves do not consist of water moving horizontally. Instead, the water is raised and lowered by the moving energy distortion. For example, a person treading water at point *a* (Fig. 10.1) would experience the energy distortion of a wave by being buoyed up as point *a* passes and lowered as point *b* passes. The person is buoyed up and then down but does not move appreciably in a horizontal direction. A surfer rides the crest of a wave, which is the energy of the wave traveling through water to the shore. Here, however, the friction of the beach in the shallow water causes the bottom part of the wave to lag, and the top part rolls over.

The activities in Experiment 10–1 will help you visualize the concept of energy transfer of a wave in a spring. You will place distortions on

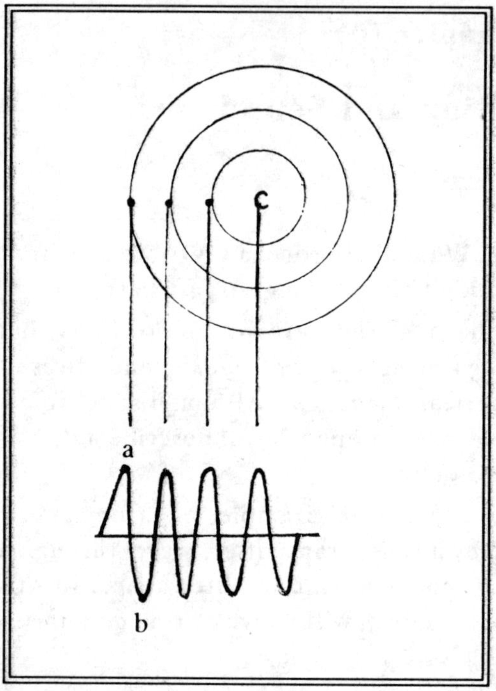

Fig. 10.1. Concentric waves radiate from *c*, the point of disturbance.

a large-diameter spring and observe their behavior as they travel back and forth.

*****

Experiment 10–1. Wave Motion

With your partner holding one end to the floor, stretch a large-diameter spring along the floor as far as possible without exceeding its stretch point, say about 6–7 meters. Now, give the spring a sharp flip to the right and observe what happens to the disturbance (wave) (Fig. 10.2a). Observe that the individual spirals of the spring do not travel, but are merely displaced laterally, just as a water wave passing a floating cork causes the it to bob up and down without moving horizontally. Careful scrutiny indicates that the leading edge of the wave causes the spring to constantly displace, while the trailing edge is constantly returning back to rest. What does the wave look like after it is reflected (bounces back)? Can you draw the wave shape and position after reflection? Now flip a series of waves on the spring. Is it inverted or reflected right-side up? Does the leading or trailing edge proceed the other after reflection? Draw the reflected wave.

Fasten a heavy spring to the large-diameter spring at the end your partner is holding. Stretch the springs along the floor for a distance of 10 m or so. Flip the large-diameter spring (Fig. 10.2b) and notice the wave at the junction of the large-diameter spring and the heavy spring. Is some of the wave transmitted? Reflected? Draw their shapes. Try sending a wave along the heavy spring toward the large-diameter spring (Fig. 10.2c) and observe the effect at the junction.

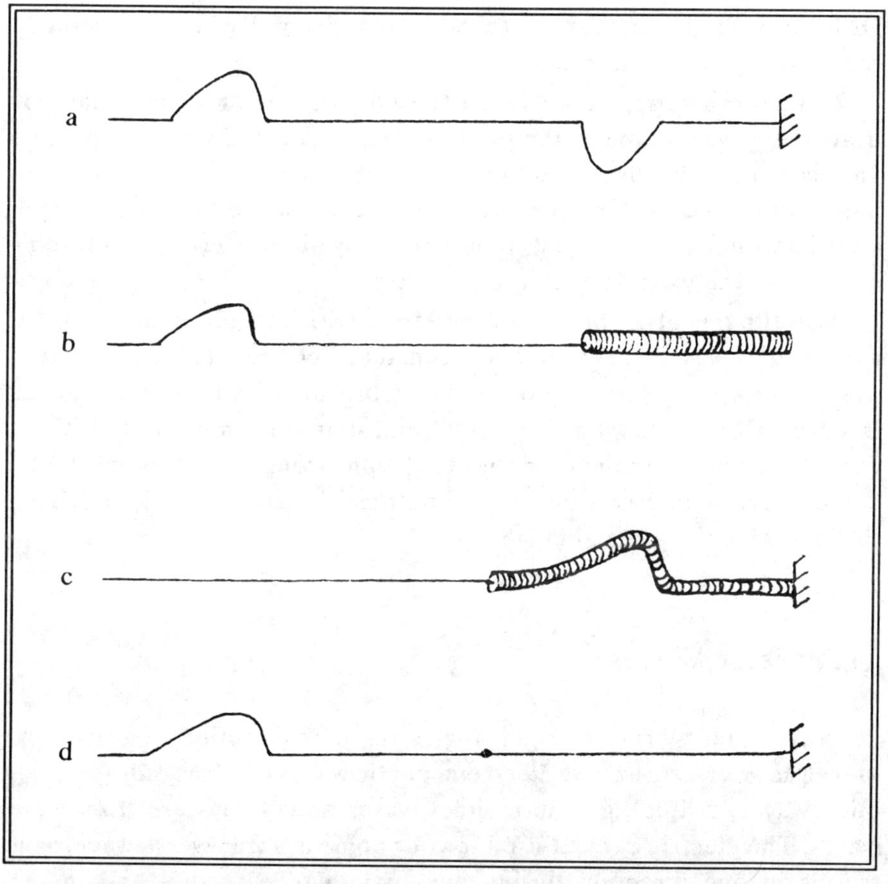

Fig. 10.2. Representations of the waves created in Experiment 10–1.

Replace the heavy spring with a 5-m length of string. Flip the large-diameter spring and observe the wave at the junction of the light string (Fig. 10.2d). Draw the shape and direction of the wave.

Remove the string from the large-diameter spring and, with your partner holding down the opposite end of the extended spring, hold your end tightly to the floor. Reach out at arm's length and grasp the spring, pulling the spirals back to your hand, in essence compressing this section of the spring. While continuing to hold down the ends of the large-diameter spring, release the compressed section and observe the wave form, direction of disturbance and direction of travel. Is this the same wave form as for the previous waves? How is it similar? Different? Try it with the stretched heavy spring. Does the tension on the spring make a difference.

Now, flip a series of waves on the large-diameter spring. Can you draw the wave's shape at the point where a reflected wave encounters an incoming wave? Note that cancellation occurs at the point the two pass over each other. In other words, the crest part of the wave neutralizes the trough part. At other points you may observe crests reinforcing crests with the formation of a larger wave.

Now flip the large spring back and forth with a regular motion so that a series of waves are generated at a constant frequency to form a pattern of *standing waves*. Notc the points of reinforcement and where the spring is at rest. Reflected waves are now reinforcing incoming waves. What can you say about the net amplitude (height from the rest point)? You can observe such waves on vertical automobile antennas while driving through winds of a certain velocity.

*****

Transverse Waves

Waves that vibrate at right angles to their direction of travel are known as *transverse waves*. Electromagnetic waves (such as radio waves), microwaves, visible light, ultraviolet waves and X-rays are transverse waves. The electric current used in your home is a transverse wave, as it travels in one direction during one-sixtieth of a second and in the opposite direction for another one-sixtieth of a second.

Figure 10.3 is a displacement-time graph of a transverse wave such as that in household electrical current. The horizontal line represents zero displacement (current flow). The displacement above zero represents current flowing in one direction, which goes to a "positive" peak and slowly decreases to zero. The current reverses and increases to a peak in the opposite direction and decreases to zero. The process reverses itself sixty times each second of time.

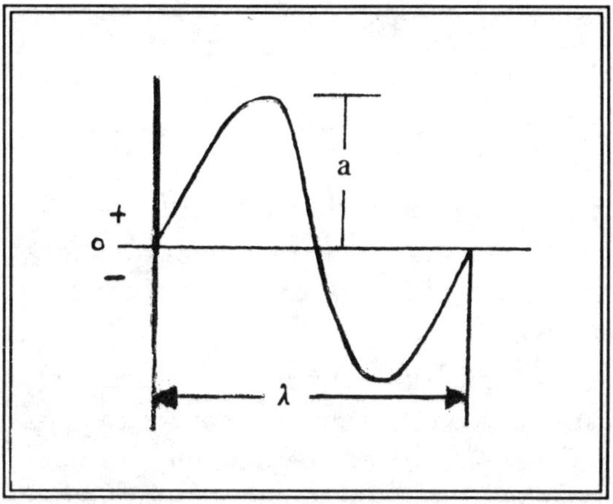

Fig. 10.3. Graph of a transverse wave.

The maximum displacement from zero is the *amplitude* (a), and the length of one complete cycle (vibration) is the *wavelength* (λ). The number of cycles that pass a given point or take place per unit of time is known as the *frequency* .

Frequency is often expressed in *hertz* (hz) units, in honor of the German physicist Heinrich Hertz and and is heard by the human ear as *pitch*. The amplitude is heard by the ear as loudness.

Frequency and wavelength are inversely related. As the frequency (vibration/time) increases, the wavelength decreases. If the frequency is doubled, the wavelength is reduced by one-half. Conversely, if the wavelength is doubled, the frequency is halved. Because frequency can be thought of as the number of waves passing a given point in one unit of time (#/sec) and the wavelength is the length of the one wave, the product of the frequency and the wavelength is the speed of the wave:

$$\frac{waves}{sec} \times \frac{m}{wave} = \frac{m}{sec}$$

Low frequency electromagnetic waves, such as those discussed in Chapter 13, need an electric conductor for travel. If the frequency is increased sufficiently, the waves are able to break away from the conductor and travel through space while sustaining themselves. Radio waves radiating from a transmitter are an example of this phenomenon. These waves do not require a medium of travel.

Longitudinal Waves

Longitudinal waves vibrate parallel to their direction of travel. This was the type of wave you formed by compressing a section of the large diameter spring and releasing it. The resultant wave was one of compressions and expansions (*rarefactions*).

Sound waves are an example of longitudinal waves. They require a medium of travel such as air. In other words, something needs to be set into vibration. Figure 10.4 is a representation of a percussion instrument setting air into vibration, producing longitudinal waves. As the diaphram of the instrument swings out, the air is compressed (a) and when the diaphram swings in, the air is rarefied.

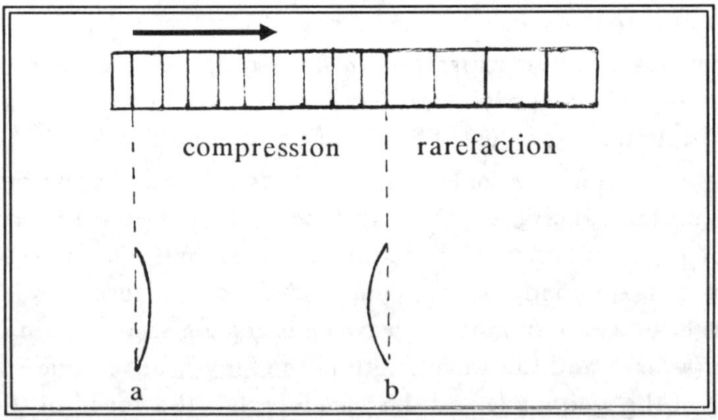

Fig. 10.4. Compressions and rarefactions from a percussion instrument.

*****

Experiment 10–2. Sound From a Tuning Fork

Strike a tuning fork against a resilient surface such as the sole of your shoe. Bring the vibrating fork lightly against your cheek. What do you feel? Strike the fork again and place the handle end on the table top or on a chalkboard. Notice the pitch is amplified. When the chalkboard or table top is set into vibration, it causes more air to be set into vibration. Strike the fork again and place the vibrating tines vertically into the surface of water in a half-filled beaker. What do you observe?

*****

Experiment 10–3. Vibrating Strings

You will need one wide and one thin rubber band, a 30-cm ruler and a pen or pencil. Hook the end of a rubber band onto the knob of a desk drawer or some other convenient, fixed point. Stretch the rubber band slightly, pluck it and note the tone. Continue plucking the rubber band as you stretch it. Now, shorten the rubber band as you pluck it. What happens to the pitch (frequency)? Fasten both rubber bands longitudinally to a 30-cm ruler with a pencil under it, as shown in Figure 10.5. Roll the pencil 10 cm from one end of the ruler and pluck the long

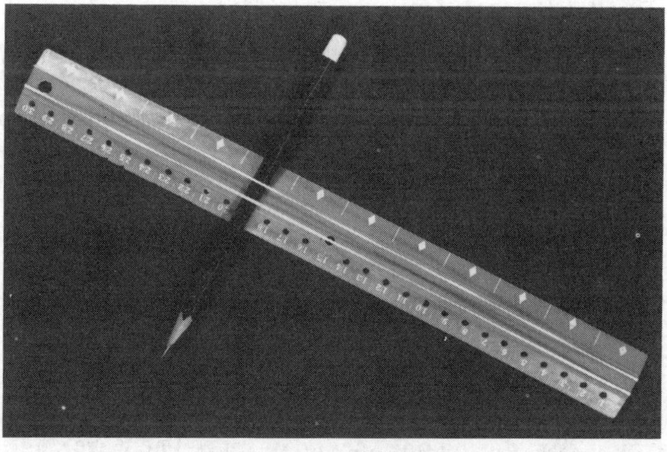

Fig. 10.5. Rubber band musical instrument.

ends of the rubber bands. Note the tones. Which has the higher tone? Continue plucking the rubber bands as you progressively roll the pencil toward your fingers, essentially shortening the rubber bands. How does shortening or lengthening the rubber bands affect the tone of the vibrating bands?

If you pluck the thinnest string on a guitar or some other stringed instrument and then pluck the thickest string, you will note the lower pitch of the thicker string. One can readily see that increasing the thickness of a vibrating string lowers the pitch and decreasing it raises the pitch. If you pluck the thinnest string while firmly holding it down at the third fret and then release the string and pluck it again, you will note a lowering of pitch. Pressing the string against the third fret is essentially shortening the string. Increasing the length of a vibrating string lowers the pitch and decreasing its length raises the pitch. If the tension on a guitar or piano string is increased, the pitch increases and if the tension is decreased, the pitch decreases. Therefore, the pitch of a vibrating string is inversely proportional to the length and thickness, but directly proportional to the tension.

Doppler Effect

As you stand along a highway listening to an automobile approaching at a high rate of speed, the sound of the tires on the highway steadily raises in pitch. As the automobile passes you and recedes into the distance, the pitch of the tires steadily lowers. This apparent change in pitch is due to the movement of a body and is called the *Doppler Effect,* in honor of the Austrian physicist, Christian Doppler (1803–1853). Doppler was able to predict the relative frequency of sound mathematically, when given the relative velocity of the listener in respect to the sound source.

An example of calculating the Doppler Effect of sound can be aided by Figure 10.6. An observer (O) on a highway hears a truck moving away at 27 m/sec sound its 700 hz airhorn for one second of time. Considering the velocity of sound as 330 m/sec, the first pulse of the airhorn travels 330 m. The last pulse travels 357 m (330 m + 27 m) because the truck has traveled 27 m during one second of time. The wavelength , λ, of the

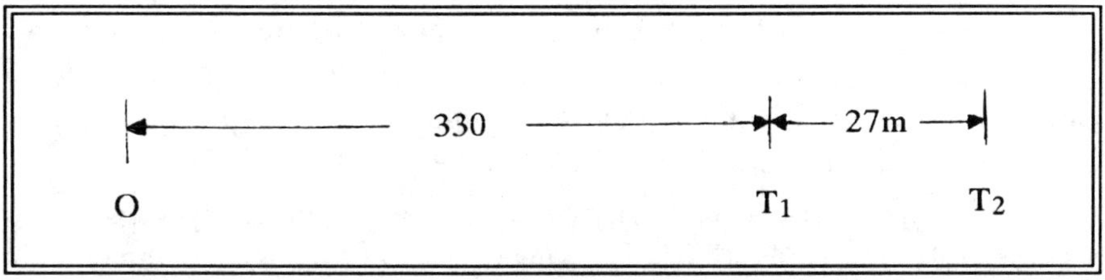

Fig. 10.6. Doppler Effect of a truck's air horn. The airhorn begins at T_1. One second later the truck, now at T_2, has traveled 27 m. The final pulse now has to travel 27 m further than the first pulse, thereby stretching the wavelength.

of the last pulse can be found by dividing the distance (*d*) by the number of pulses:

$$\lambda = \frac{357 \text{ m}}{700 \text{ pulses}} = 0.51\text{m}$$

To compare the difference in frequency (pitch) of the first pulse to the frequency of the 700th pulse, consider the relationship of frequency (*f*), time (*t*), wavelength (λ) and velocity (*v*). Because velocity is the length of one wave divided by the time:

$$v = \frac{\lambda}{t} \quad \text{and} \quad t = \frac{\lambda}{v}$$

The frequency (f) $= \frac{1}{t}$ and t $= \frac{1}{f}$.

Substituting this expression for *t* into the velocity formula yields:

$$\frac{1}{f} = \frac{\lambda}{v} \quad \text{or} \quad f = \frac{v}{\lambda}$$

The frequency of the last pulse can now be calculated by substituting the value of its wavelength into the frequency formula:

$$f = \frac{v}{\lambda} = \frac{330 \; m/sec}{0.51 \; m} = 647 \; hz$$

At this instant in time, the frequency heard by the observer is 647 *hz*, compared to the 700 hz heard from the truck initially. Thus, a marked decrease in tone is heard. If the truck was approaching the observer, a marked increase in frequency could be calculated.

The Doppler Effect has many practical uses. Certain types of radar use the shift in wavelength of radio waves reflected from a moving body to determine the body's speed. Sonar also can detect the speed of a moving body by the Doppler Effect of the body. Astronomers use the Doppler shift from the light of stars to determine whether a star is receding or approaching. As shall be seen later, the color of light is a function of its wavelength.

Science/Technology/Society

Science and technology have done much to improve peoples' enjoyment and use of sound. Auditoriums, concert halls and outdoor stages have been engineered to project the maximum quality and volume of sound. The Morman Tabernacle in Salt Lake City, Utah, is a tribute to such engineering excellence.

Sound can be used to alter radio waves so they can be transmitted great distances and manipulated to reproduce the original sound with remarkable fidelity. Stereophonic FM radios and televisions allow people, even in remote areas, to enjoy quality music, often via satellite transmission.

Recording processes allow storage and access to sound long after the original was produced. Again, the quality of reproduction has been enhanced to the point of remarkable fidelity. Warning devices, such as fire sirens, ambulance warblers, police sirens and weather-alert devices, save lives and prevent accidents. The list is enormous, and individuals have come to depend greatly on the use of sound.

But all the advancements in sound technology can be detrimental to mankind. Powerful tape recorders and portable radios now pollute the peace and quiet of streets, beaches and remote areas. One may find the quiet of even a remote campsite penetrated by a loud throbbing cacophony. A rise in pulse rate can be noted during a quiet evening walk when an automobile whizzes by with the stereophonic tape-player or radio blaring at seemingly full volume.

The very devices designed to save lives, such as sirens and the horns of automobiles, blend into a city background, grating on one's nerves. There is mounting evidence that heavy amounts of background noise can have a negative effect on the psychological and physiological states of people. Nature seems to set at least one example of a solution for noise pollution. Planting heavy vegetation along freeways, busy streets and railroads absorbs much of the noise. The use of sound-absorbent materials along busy streets also can help.

Citizens need to be informed and consider all of the implications of making changes in the environment. Care must be taken in the selection of sites for shopping centers, freeways, boulevards, overhead railways and airports, and choices must be made with the welfare of the entire community in mind.

Even those living in agricultural areas are not immune to noise pollution. Farmers and factory workers are subject to heavy noise from tractors, loaders and other machines. It is now not uncommon to see farmers, construction and assembly-line workers wearing sound-muffling devices over their ears during the operation of machinery.

Perhaps one of the most ominous, yet subtle, forms of noise pollution is that from the entertainment industry. The level of sound produced electronically is now amplified well above the point where permanent hearing damage can occur. Young people, especially, subject their hearing to rock concerts, compact audio disk players and radio/tape player headsets. Close-fitting headphones allow dangerously loud sounds to be applied directly to the ears! The time is approaching when hearing aids may become standard equipment for people in the 20-year-old age group!

Sound is very beneficial to mankind. To appreciate how much you depend on sound, you could wear earplugs for just one day.

CAUTION: Do not try this while driving a vehicle or doing any type of hazardous activity!

Chapter 11

Light

You are constantly surrounded by light. Even during a moonless night, a certain amount of light exists in the sky and the dark background of the night sky is punctuated by the apparent twinkle of stars. The phenomenon of light is critical to many aspects of human endeavor; thus, an understanding of light is of considerable value.

Some Characteristics of Light

For centuries people contemplated what light was. Of special interest was whether light had speed or was instantaneous. Galileo, who used experimentation to verify scientific claims, attempted to measure the speed of light by having an assistant uncover a distant lantern and observing the length of time it took to become visible. But light was too fast to be measured by this process. Olaf Roemer (1644–1710), a Danish astronomer, attempted to measure the speed of light by calculating the difference in the time it took light to pass across the earth's orbit after being reflected from one of the moons of Jupiter. Although not successful, he did demonstrate that light has a finite speed.

Seventeenth-century scientists argued about whether light was a wave or consisted of particles. Christian Huygens (1629–1695), a Dutch physicist-mathematician, argued for the wave theory of light, but could not explain the photoelectric effect using this model. Isaac Newton took the position that light consisted of particles. He was able to demonstrate the efficacy of his model mathematically, but he could not adequately explain light bending around the edges of objects nor convincingly explain refraction.

The present model of light incorporates the dual nature of light, which includes both the particle theory and the wave theory. This model is useful in explaining how light is bent in passing through narrow slits and the photoelectric effect. Your study of light will examine several of the characteristics of light that led to the presently accepted model.

Straight-Line Propagation

A fundamental assumption about light is that it travels in straight lines. You cannot see around corners with the unaided eye, but light can be bent and reflected in non-linear paths. In the following discussions of light, observed phenomena will be explained by tracing the travel of light with straight lines called *rays*. You should not assume that light travels in rays, but realize that the term is used for convenience in describing the paths in which light travels.

Parallax

While driving along a highway, you may have noticed that the power-line poles appear to be moving more rapidly than the trees in the background. Of course, you realize that you are the one moving, and therefore the motion between the poles and the distant trees is an illusion. This apparent relative motion is known as *parallax*.

You are going to do several experiments using parallax to locate images. To conceptualize this process, place one index finger in front of your eyes at arm's length and in a vertical position. Now, hold up the index finger on your opposite hand directly in front of your eyes and move your head from side to side. Both fingers appear to move in respect to distant objects, but the closer finger appears to move faster than the more distant one. As you move the nearer finger closer to the distant finger, parallax decreases until it is eliminated when the fingers touch. Parallax can only exist if two objects are separated. For example, you can locate the image of a pencil in a plane mirror using parallax. Place one pencil in front of a mirror and another pencil behind the mirror. While moving your head from side to side and observing the image of the first pencil in the mirror, move the pencil behind the mirror until parallax is eliminated.

Reflection

Light can only be seen by humans either directly or by reflection. For example, you can see a shaft of light passing through a darkened room by either looking directly at the source or by seeing it reflected by dust particles or some other object in the room.

Reflection is the phenomenon of light bouncing off the surface of an object, thus changing its direction. If your eye is in the path of the reflected light, you will "see" it. Because light is believed to travel in straight lines, it is possible to trace its path in the formation of images.

*****

Experiment 11–1. Reflection From a Plane Mirror

Draw a line bisecting the length of a piece of 8 1/2 x 11 inch white paper. Place the paper on a piece of cardboard. Support the mirror vertically with modeling clay such that the silver coating on the back of the mirror is directly over the line on the paper. Outline the mirror with a pencil..

Stick a pin, which you will use as an object to view in the mirror (object pin), about 10 cm directly in front of the mirror. Stick a second pin somewhere near the front-left corner of the paper. Stick a third pin in the paper near the front of the mirror such that it lines up exactly with the second pin and the image of the object pin. Repeat the process near the front-right corner of the paper.

Remove all of the pins and stand a nail on its head exactly where the object pin was located. This will be the object nail. Place a second nail on its head behind the mirror where you estimate the image to be located. Using parallax, locate the nail exactly where the image of the object nail is located. Carefully mark this spot and label it *image*. Also be sure to label the spot where the object is located.

Align a straight edge along the two pin holes that line up with the object image and draw a solid, straight line up to the mirror-line representing the silvered surface of the mirror. Continue with a dashed line behind the mirror to the paper's edge. Draw a solid, straight line from the point where the line you just drew intersected the mirror-line to the object pin hole. Repeat the process with the two pin holes on the opposite side.

The point where the two dashed lines intersect behind the mirror is where the image should be located. How does this point compare with the point established by parallax? Which point is more realistic? How can you tell?

Carefully measure the distance from the object to the mirror and the image to the mirror. How do the distances compare? Can you formulate a rule for objects and images in plane mirrors? The light from the object pin moved to the mirror and is reflected back to your eye (Fig. 11.1). Because light travels in a straight line, you are deluded into thinking the light from the pin comes from behind the mirror, so the image appears to be behind the mirror. Such an image is a *virtual image.* It is not real and exists only in your mind. If you look behind the mirror, the image is not there. Unlike real images, virtual images cannot be projected on a screen.

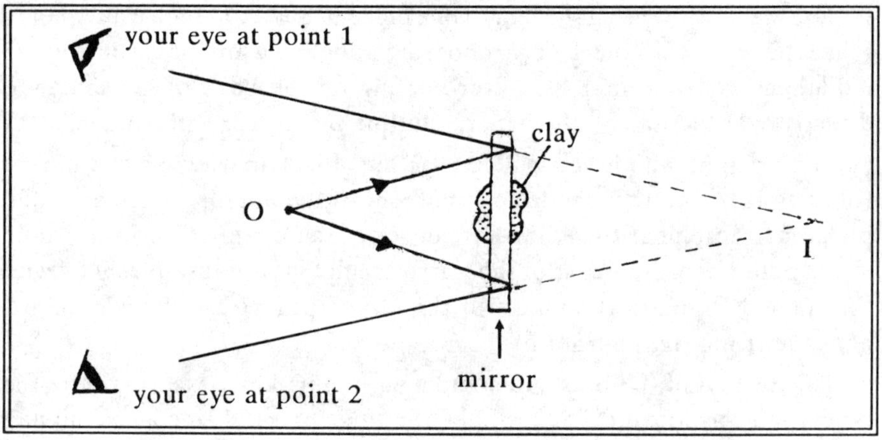

Fig. 11.1. Locating the image in a plane mirror.

Draw a line perpendicular to the mirror (normal) at each point where the line from the object intersects the mirror-line. At both points, measure the incoming angle (*angle of incidence*) and the normal line and the reflected angle (*angle of reflection*) and the normal line. How do they compare? Can you state a general law of reflection by comparing the angle of incidence and the angle of reflection?

Concave Mirrors

*****

Experiment 11–2. Images of a Concave Mirror

Mount a concave mirror vertically and a battery and bulb on a strip of masking tape on a table as shown in Figure 11.2. Begin with the object bulb near the end of the tape strip. Locate the image using a small piece

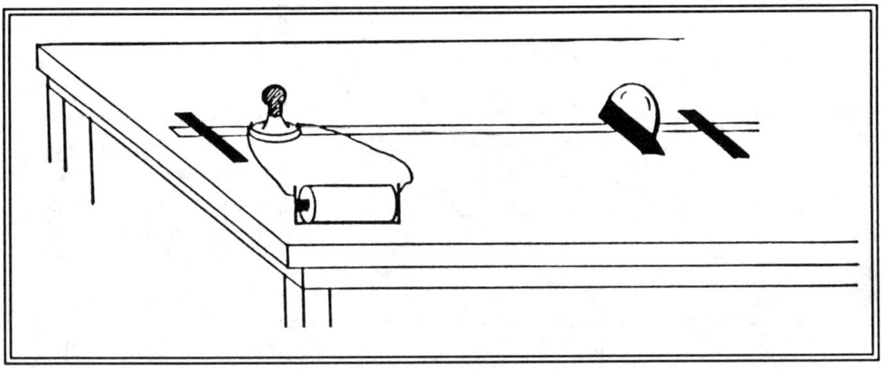

Fig. 11.2. Concave mirror apparatus.

of white cardboard. Is the image upside down or right side up? Is it real or virtual? Is it larger or smaller than the object? Can you locate it by parallax? How does the position located by parallax compare to the position located by the image-card?

Move the bulb toward the mirror in discrete steps. Mark both the object and image positions and label them. Continue this process until the image goes off the end of the tape. How does the size of the object compare to the size of the image at each step? Try to find the points where the image is the same size as the object and about twice as large as the object and mark it on the tape. Is the image in each case real or virtual? Where is the object located in each case?

As you approach the mirror, you reach a point where the image becomes so large you will not be able to find it because it becomes a large blur. At this point you have reached the *principal focus*. This point is where all the light from a very distant object, say the sun, is reflected to a point (Fig. 11.3). Light from the sun would be so concentrated here that you could ignite a piece of paper, or, if your mirror was large enough, melt steel! Because light rays coming in parallel to concave mirrors always converge on the principal focus, they are known as *converging mirrors*.

Now, move the bulb between the mirror and the principal focus and try to locate the image by looking into the mirror. The image should appear right-side up. Is it real or virtual? Is it enlarged or smaller than the object?

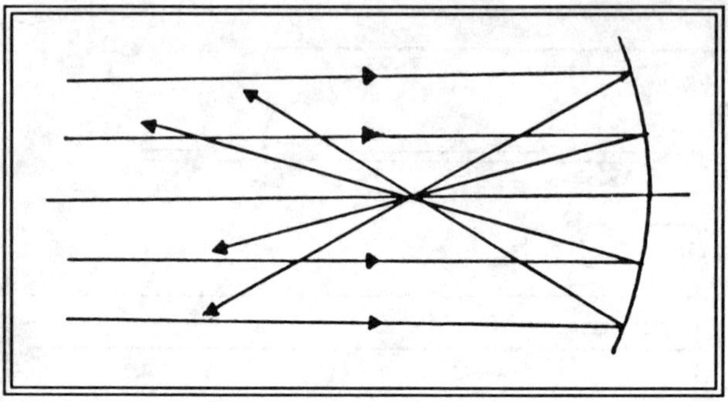

Fig. 11.3. Parallel light rays incident upon a concave mirror are reflected back through the principal focus (f).

*****

The observations you made in Experiment 11–2 can be explained theoretically by line drawings such as those in Figure 11.4. Arrows are used for objects because they distinguish up from down.

Objects located at various positions along the principle axis of a concave mirror form various-sized images, both real and virtual (Fig. 11.4). Lines representing imaginary rays are drawn to indicate the paths taken by the light. The intersection of at least two rays form an image. For convenience sake, only one point on each object is used in drawing each image, but try to imagine an infinite number of rays coming from each object to form its image.

Light coming in from long distances (Fig. 11.4a) impinges upon the mirror essentially parallel to the principal axis and is focused at the principal focus (f). Reflecting telescopes use this arrangement for gathering light from the stars. The real image at the focal point can be photographed or viewed directly through a magnifying apparatus.

As the object is moved closer to f, the image becomes larger and is observed to be inverted. With the object at a finite distance from $2f$, the image is real, inverted, smaller than the object and between $2f$ and f (Fig. 11.4b). At $2f$ the image is the same size as the object, inverted and real (Fig. 11.4c).

When the object is between $2f$ and f, the image is inverted, real and larger than the object (Fig. 11.4c). At the focal point the image becomes infinitely large (Fig. 11.4d), such that it becomes blurred. Between f and the mirror, the object has an image which is upright, virtual and larger than itself (Fig. 11.4e).

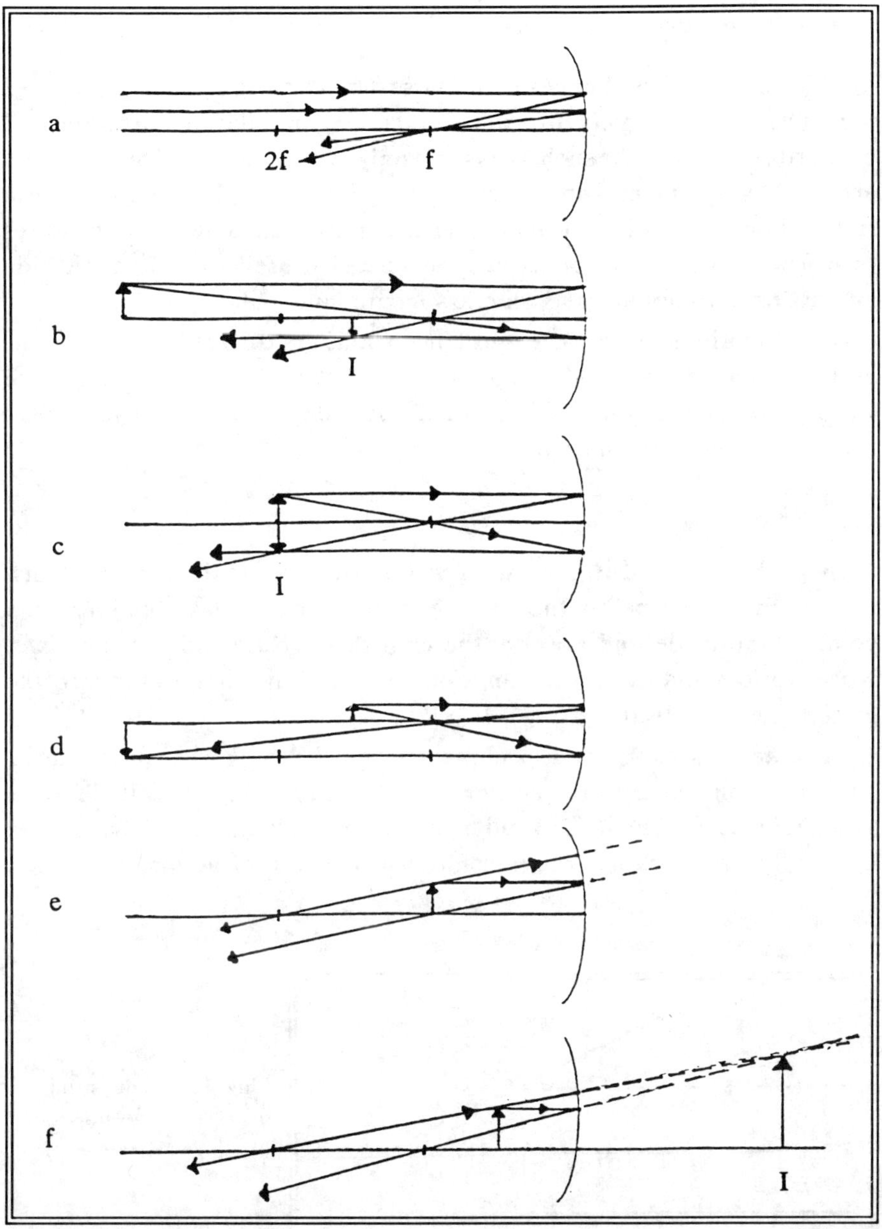

Fig. 11.4. Images formed by a concave mirror.

Convex Mirrors

Whereas you found concave mirrors to be converging mirrors, convex mirrors are diverging mirrors, in that incoming parallel rays are reflected outward (Fig. 11.5). As such, they form only virtual images. These mirrors are used as ornamental spherical garden mirrors and, because they form reduced images, they are used for security purposes in department stores to provide a wide field of view. They are also used as wide-angle side mirrors on automobiles and rearview mirrors on boats.

Combinations of concave and convex mirrors distort images of your body at fun houses. Both magnified and reduced virtual images of various portions of your body are displayed, depending upon your location in respect to the mirror.

Refraction

A pencil inclined in a glass of water will appear disjointed or distorted (Fig. 11.6) below the surface. Set an empty glass tumbler on a coin. Stand aside and observe the coin through the side of the glass. Now, while standing in the same position, slowly pour water into the glass. The coin disappears.

You have just observed examples of light bending as it passes at an angle from one medium to another. This bending of light is called refraction. Doing Experiment 11–3 will help you conceptualize the refraction of light and a mathematical relationship that predicts refraction.

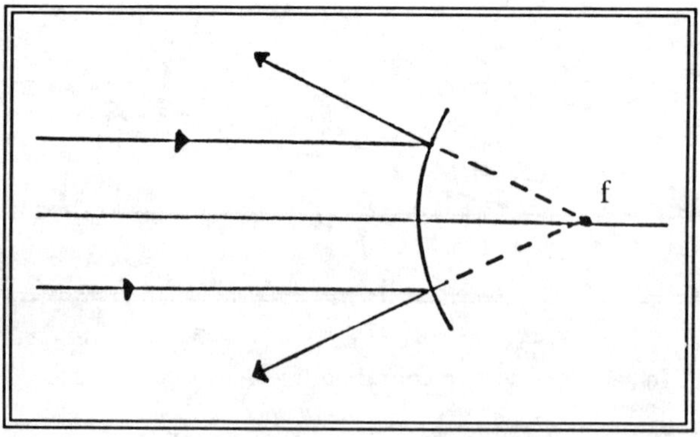

Fig. 11.5. A diverging mirror. The focal point is virtual.

Fig. 11.6. An inclined pencil in water.

*****

Experiment 11–3. Refraction of Light

To examine what happens to light as it moves at various angles from one medium to another, in this case air to water, draw a line bisecting the long dimension of a piece of paper. Place the paper on a piece of cardboard. Make a vertical mark on the center edge of the flat face of a semicircular plastic dish. Place the dish, three-fourths full of water or some other liquid, on the center of the paper such that the flat edge corresponds to the line. Make a mark on the paper directly below the vertical mark on the dish (Fig. 11.7).

Stick a vertical pin about 5 cm directly in front of the flat edge of the dish. This will be the object pin. Sight and stick a pin about 5 cm from the curved side of the dish so that it lines up exactly with the scratch on the dish and the image of the object pin. Label both the object pin points and the image pin points *#1.* Move the object pin about 10 degrees to the left and repeat the process, labeling the pin-marks *#2.* Take at least two other readings while moving the object pin progressively to the left at about 10 degree intervals. (Judge, don't measure!)

Remove the dish and pins. Draw a line through the vertical dish-mark point and perpendicular to the horizontal. Draw lines from the object pin-holes to the vertical dish-mark point and from that point through the corresponding image pin holes in front of the dish. Place a compass point on the vertical dish-mark point and strike a circle with a radius of

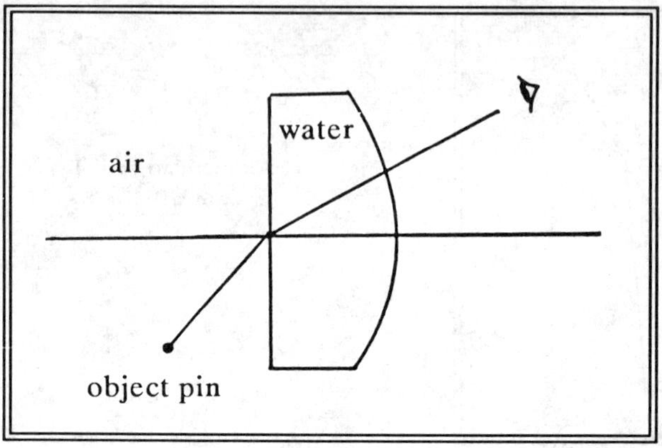

Fig. 11.7. Refraction of light arrangement.

4 cm. Draw chords from the point where each of these lines intersect the circumference of the circle (Fig. 11.8). Construct a table of the chords and the ratio of chords for each pair of rays. What do you observe about each ratio? Compare your data with the data of students using different liquids.

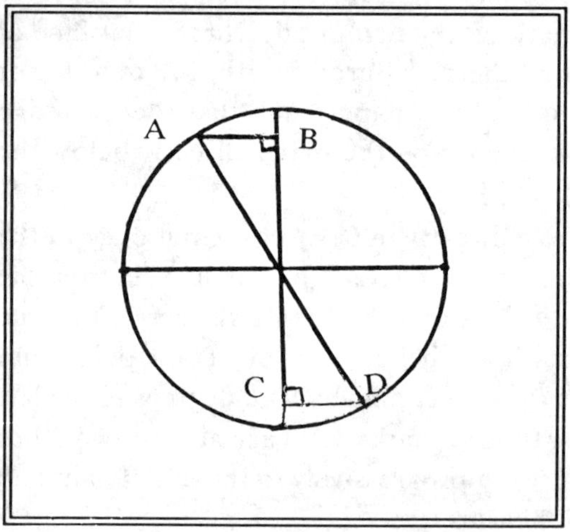

Fig. 11.8. Construction of the chords for a pair of rays.

*****

In Experiment 11–3 you observed that the ratio of the chords for a given liquid was the same for each pair of rays. Because the ratios always divided out to be the same number, they are said to be *equal to a constant.*

Because in trigonometry the definition of the *sine of an angle* is the *ratio of the opposite side to the hypotenuse* and because the radius of the circle is the hypotenuse of each triangle formed (Fig. 11.8), the *ratio of the chords is equal to the ratio of the sine of the angles.* Thus, the angles of the incident ray and the refracted ray can be measured and their respective sines looked up in a sine table to determine the ratio which is the constant for the material. This constant is known as the *index of refraction.* The relationship is referred to as *Snell's Law*, in honor of the Dutch scientist Willebrord Snell who suggested it in 1621, and is expressed as:

$$n = \frac{\sin i}{\sin r}$$

The phenomenon of refraction can be explained by the relationship where $\sin i$ and $\sin r$ are the sine of the angle of incidence and the sine of the angle of refraction respectively and n is the index of refraction:

$$n = \frac{\text{speed of light in a vacuum}}{\text{speed of light in a substance}}$$

Because light travels fastest in a vacuum, indices of refraction for substances are comparisons of the speed of light in a vacuum to the speed of light in the respective substances. Whenever a ray of light strikes the boundary of two different media (such as the boundary of air and water) at an angle, the portion of the ray at point *A* (Fig. 11.9) strikes the water first and is slowed down. The portion of the ray at point *B* continues in the air at the original speed and does not slow down until point *B* reaches the water. Thus, the portion of the ray *BC* is the length gained by this side of the ray, with the resultant bending of the ray toward the normal. If the light ray were traveling from the water into the air, the opposite would take place with the light being bent away from the normal ray (Fig. 11.9).

Lenses

Lenses are light-benders and are referred to as *refractors.* Different shapes of lenses refract light in different ways with differing results.

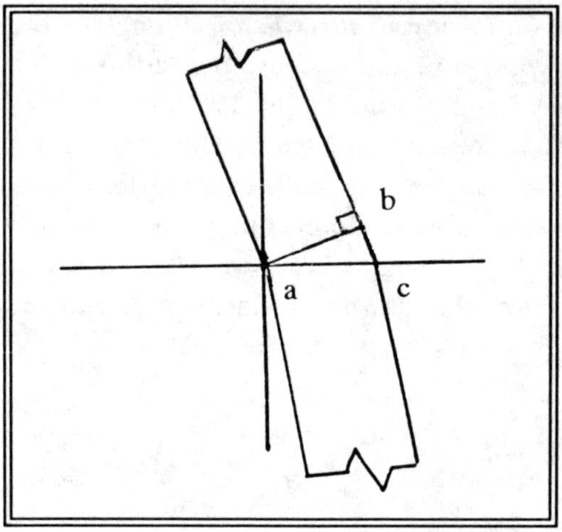

Fig. 11.9. A light ray of thickness, *AB*, traveling in the air, impinges on the water at an angle, meeting the water surface at point *A* and is slowed down. Portion *B-C* of the ray continues in the air until point *B* strikes the water surface where it is slowed down. Thus, point *B* advances ahead of *A* and the net result is the ray is bent toward the perpendicular.

*****

Experiment 11–4. The Converging Lens

Place a lighted bulb and socket at one end of a long strip of masking tape as shown in Figure 11.10. Looking through the opposite side of a convex lens, locate the image by parallax. Move the bulb closer to the lens in steps, noting the position of the object (bulb) and the image as well as its relative size. Is the image upright or inverted? Is it real or virtual? How does its size compare to the object? Where is the image about the same size as the object? Mark the spot. Hold the object

Fig. 11.10. Converging lens arrangement.

several meters away from the lens. Locate the image (You may need to use a small paper screen). Mark the spot.

Move the object next to the lens. What kind of image do you have? Describe it. Slowly move the object away from the lens, locating the image at various intervals. Where does the image get so large it becomes distorted?

Where is the principal focus of the lens? Can you demonstrate similar characteristics working from the opposite side of the lens? Does the lens have two principal foci? Can you demonstrate them?

What generalizations can you make about a converging lens? Can you see parallelism with a converging mirror?

*****

Your observations in Experiment 11–4 can be explained through the line drawings in Figure 11.11. Parallel rays incident upon a converging lens will be refracted to a point on the opposite side of the lens (Fig. 11.11a). This point is called the *principal focus*. Light rays coming from a distant object, such as a star, will be parallel. The image of the star will be very small, and it will be located at the principal focus. An example of a lens used in this configuration is the refracting telescope.

When the object is a finite distance beyond $2f$ (Fig. 11.11b), the image is inverted, real, smaller than the object and located between f and $2f$. The real and inverted image becomes the same size as the object when the object is at $2f$ (Fig. 11.11c). Between $2f$ and f the object produces a magnified, inverted, real image (Fig. 11.11d). An example of this is the filmstrip projector, which produces an image that is larger than the object (filmstrip) and is upright and virtual with the object between f and the lens (Fig.11.11e). When the object is on the principal focus (f), the image becomes infinitely large and blurs. Placing the object at any point between f and the lens yields an image that is virtual, larger than the object and appears behind the object.

Similar relations between the object and the image exist if the object is placed on the opposite side of the lens because the lens also has a principal focus on this side. Converging lenses are used as magnifying glasses, in telescopes, projectors, cameras and eye glasses that correct far-sightedness.

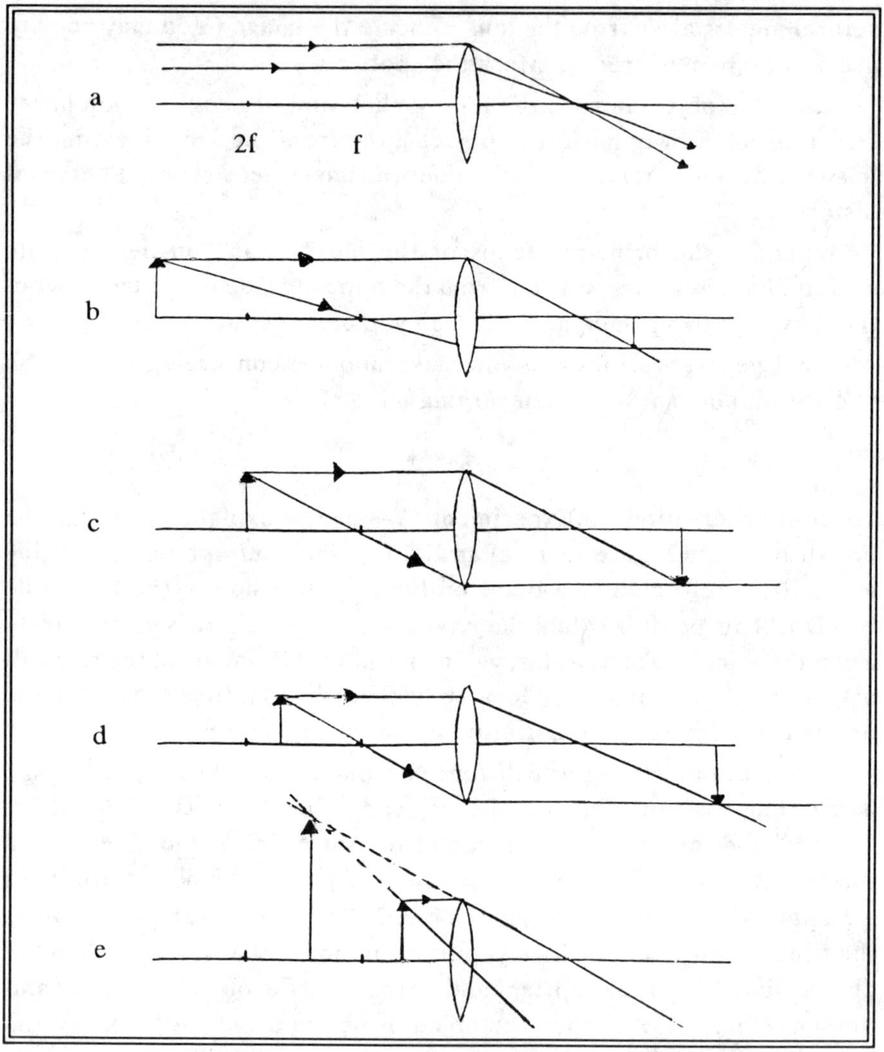

Fig. 11.11. Image-object relationships for a converging lens.

The Diverging Lens

A diverging lens is constructed thicker at the edges to refract light away from the principal focus (Fig. 11.12). Such lenses only form virtual images, and are used to correct distortions in lens systems and eye defects such as nearsightedness.

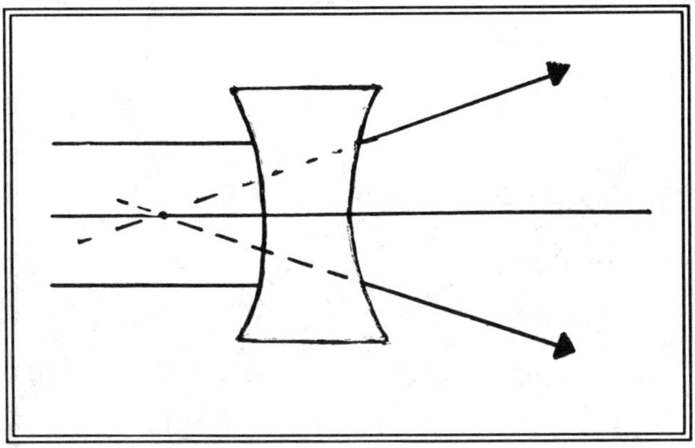

Fig. 11.12. Parallel light rays through a diverging lens.

Spectra

Newton observed that sunlight passing through a glass prism produced the colors of the rainbow. He also observed that a white light was produced when a second prism was placed in the path of the rainbow colors emanating from the first prism, and therefore, reasoned that white light is a combination of all colors. Further analysis shows that each color in the white light is refracted at an angle inversely proportional to its wavelength (Fig. 11.13). Therefore, the violets and blues are refracted more than the reds, oranges and yellows. The colors are spread out into a pattern called the spectrum.

This phenomenon can be observed in nature. A rainbow appears in the sky only when rain is occurring and the sun is shining from behind the viewer. Color dispersion takes place because the different wavelengths of the light are refracted differently, just as with a glass prism. Sunrays enter the raindrops and are refracted, reflected and refracted again on the way out of the drop. Only that light that leaves certain drops, those that make a specific angle between the incoming sunray and the ray that reaches your eye, makes up the rainbow. Thus, the rainbow you see may not be exactly the same rainbow seen by another person!

Color dispersion is also observed on the surface of soap bubbles and oil films on water. Some of the light is reflected from the top surface of such films, while a portion of it is transmitted to the opposite side of the film and reflected back. Hence, the transmitted light travels an extra distance equal to two thicknesses of the film. If the film is of the correct

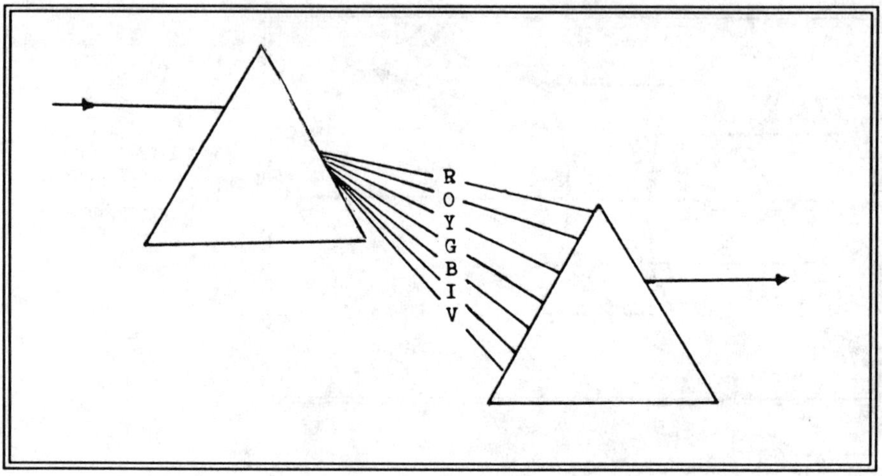

Fig. 11.13. Sunlight traveling through two prisms.

thickness, the crest of the reflected wave reinforces the crest of the transmitted wave and a bright band results (Fig. 11.14a). Different colors have different wavelengths, and reinforcement will take place with different thicknesses of films. Cancellation of light takes place when a crest of the reflected wave coincides with the trough of another wave (Fig. 11.14b) and results in a dark band. Because a thin film varies in its overall thickness, you usually observe the effect of white light being reflected from the two surfaces of the film alternated with dark bands and bands of different colors of the spectrum.

*****

Experiment 11–5. Flame Tests

CAUTION: Goggles required for this activity.

For this experiment you will need vials of calcium chloride, copper chloride, copper nitrate, lithium chloride, potassium chloride, sodium-hydrogen carbonate, sodium chloride, strontium chloride, strontium nitrate and ten balsa sticks with nichrome loops. Record the names of the compounds. Heat a loop in the alcohol burner until it glows and then quickly dip it into one of the compounds.

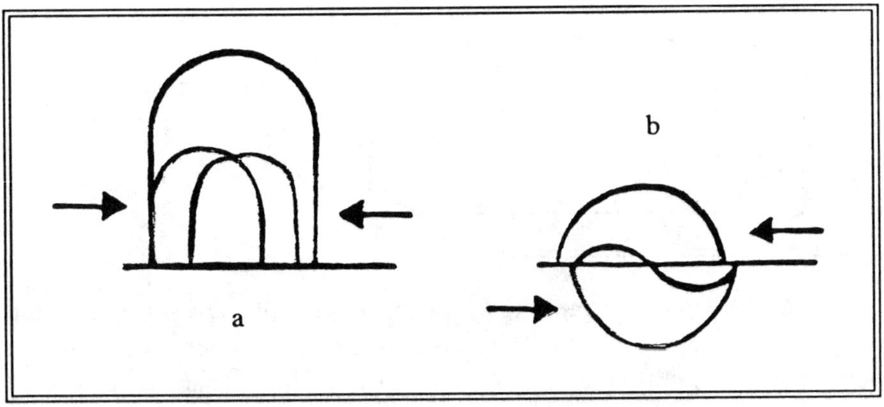

Fig. 11.14. (a) Bright bands caused by the reinforcement of two waves, and (b) dark bands formed by an incoming and reflected wave.

Place the wire with the bead of compound in the bottom edge of the flame so no drips fall into the flame. If some material falls onto the wick of the burner, the wick will become contaminated and the characteristic color of the contaminant may persist throughout the experiment, making discrimination difficult if not impossible. Observe and record the color of light emanating from the heated compound.

Can you identify any patterns? Can you identify the ions responsible for specific colors? Have you identified some characteristic properties?

The characteristic color given off by each respective element is due to the excitation of the electrons in their outermost energy levels. As the elements are heated, their outer electrons gain energy. As these electrons lose their energy, it appears in the form of light, with each color representing a different energy level.

*****

Experiment 11–6. Diffraction of Light

Coat a microscope slide with a colloidal suspension of graphite. After the graphite has dried, use a single-edged razor blade (or a needle point) to cut a single slit in the graphite near the end of the slide. Use another slide as a straight edge. About 2 cm from the opposite end make about six slits very close together (Fig. 11.15).

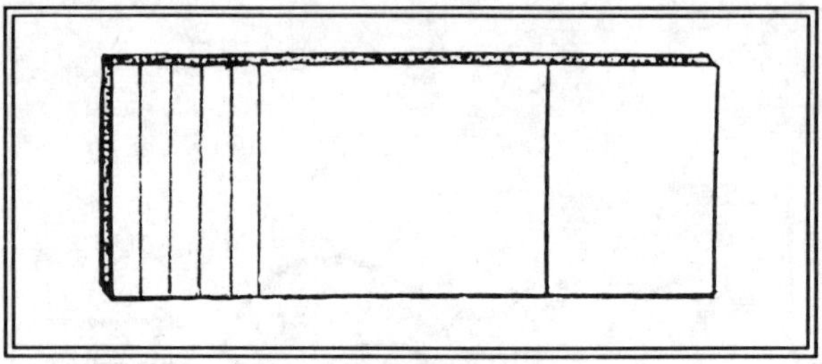

Fig. 11.15. Slits cut in the graphite coating on a microscope slide.

Look at a showcase lamp through the single slit. Do you notice the vertical bars? Observe the colors of the spectrum on either side of the slit. Actually, you can observe the same effect by looking through a slit formed by holding two adjacent fingers of your hand very close together. Now look at the showcase lamp through the multiple slits of the slide. Now look at a mercury vapor light. The patterns you observe are *diffraction patterns*.

The width of the slits you constructed are considerably wider than the wavelengths of light. Therefore, you can consider light coming through the slit as several wavelets (Fig. 11.16). As these wavelets radiate out from the slits, they interfere with each other. As a result, reinforcing crests cause bright bars and reinforcing troughs cause cancellations or dark bars in the same way that crests reinforced crests and troughs reinforced troughs when you reflected periodic waves on the large spring to produce a standing wave (Experiment 10–1).

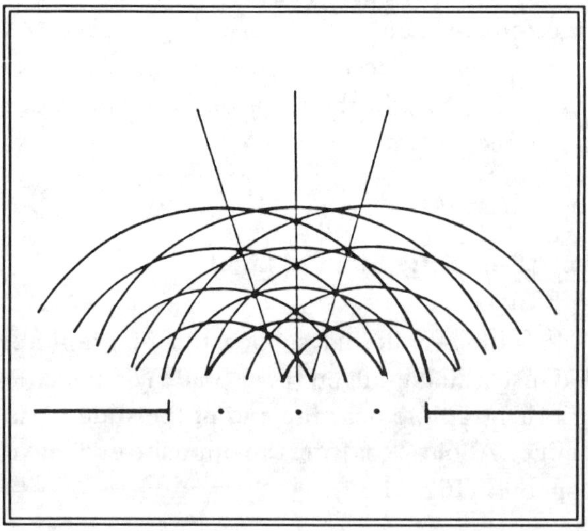

Fig. 11.16. The width of a single slit, greatly exaggerated.

*****

A similar effect can be achieved using two narrow slits placed close together, such as those constructed by drawing two single-edged razor blades across the graphite emulsion on a slide. Here, the slits act as two point-sources emitting light. The waves from the two sources reinforce each other in a line passing halfway between the two and straight ahead. Alternating light and dark bands of reinforcement and cancellation, respectively, radiate to the right and left, a similar pattern to that which you observed from the single slit.

*****

Experiment 11–7. Emission Spectra of Elements

You will now use the principle of diffraction to observe the light emission of sunlight, incandescent lamps and various excited gases. To accomplish this, you will use a piece of transparent plastic that has several thousand slits per inch, known as a *diffraction grating*, mounted in a 2 x 2 cardboard slide holder. You cannot see the slits because there are so many of them spaced very closely together. But as you move the grating at an appropriate angle in sunlight, you will notice the reflection of rainbow colors. This phenomenon can also be observed when long-play phonograph records or compact disks are placed in sunlight.

Hold the diffraction grating up to your eye and examine an ordinary light bulb. Notice the images of the spectrum. What shapes are they? Now look at a showcase lamp. What are the shapes of the images now? Look at a sodium vapor lamp through the diffraction grating and note the color and shape of the image(s). It is apparent that the shape you observe is the shape of the source. Look at electrified gas tubes of (a) sodium; (b) helium; (c) hydrogen; (d) mercury vapor and (e) a fluorescent light tube, tubes glowing as a result of a high voltage being applied. Note the order of the colored lines you observe in each case. Now, see if you can identify an unknown source by looking at one of the above gases, without a label supplied by your instructor.

*****

A device that employs a diffraction grating to lay out precise patterns of light given off by materials heated to very high temperatures is called a

CAUTION: Very high voltages are used to electrify the gas tubes. Do only under the supervision of a qualified person and observe appropriate precautions.

spectroscope. Light from the substance under test is passed through a very fine slit prior to its passing throughout the diffraction grating. The images formed by the diffraction grating are very fine lines dispersed in a linear pattern. These *spectral lines* are then photographed to give a permanent record. The position of the lines can be measured. By studying the positions of the various spectral lines, the material(s) responsible for them can be identified (Fig. 11.17) and calibrated. You can also take precise measurements of the position of the lines.

Each element has its own unique set of spectral lines appearing in a definite pattern. If the light from two elements is contained in the same sample, all of the lines from both elements will appear in the spectrum, but each can be sorted out on the basis of its position. Notice the spectra of two different elements in Figure 11.18.

Fig. 11.17. Representation of the spectral lines of a gas.

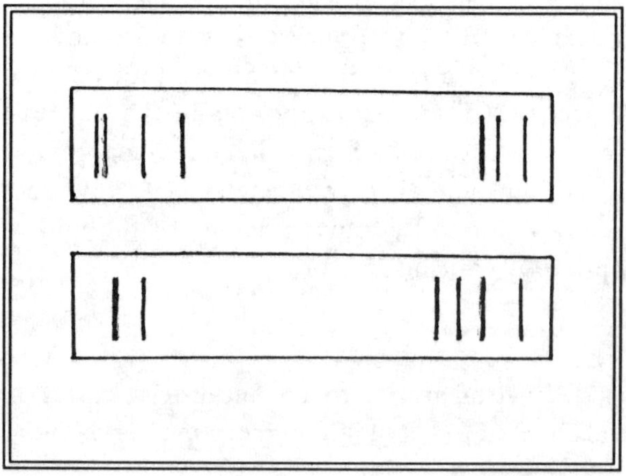

Fig. 11.18. Spectral lines representing two other gases.

Spectral analysis is a method of "finger printing" the elements, in that each element has a unique set of spectral lines, just as you possess a unique set of fingerprints. A good example of how spectral lines can be used to identify elements is the discovery of the element helium in 1869, when its spectral lines were first observed in sunlight passed through a spectroscope. Helium was identified on the surface of the earth in 1895. It was named from the Greek word for the sun, *Helios.*

Science/Technology/Society

The study of light has improved the living conditions of mankind and expanded its knowledge base. In effect, the illumination of the dark has allowed daylight to be extended. Productivity and meaningful use of leisure time have been increased, particularly during the winter months.

Through the knowledge gained from the laws of reflection, gigantic telescopes with minimum distortion have been constructed to study the sky. Mankind has found its position in the solar system. Navigation has been much improved from the study of the stars. The laws of the reflective properties of parabolic reflectors have been put to good use in microwave radio transmission.

The production of eye glasses alone has improved the human condition immensely. Consider the number of people you know who now wear eye glasses. If it were not for corrective lenses, many people could not read, do close-up work, drive an automobile or enjoy television.

The study of lenses has also led to the construction of powerful microscopes that have been instrumental in acquiring knowledge about the microscopic world. Mankind has been able to study and attack diseases through magnification of bacteria unknown to previous generations. In fact, the whole concept of health as it relates to hygiene has been largely the result of the confirmation of the existence of health-threatening micro-organisms. Magnification enables doctors to perform delicate surgery on small human organs such as the eye. Manipulation of the minute instruments and the sewing of micro-stitches would not be possible with the unaided eye.

The science of metallurgy has been greatly advanced through the use of spectral analysis, through which the purity of many substances can be verified. It can even be used in the solution of crimes by verifying the elements in soil samples, tire marks and the like.

On the negative side, the use of light has become a form of pollution or at least adds to environmental pollution. Night lights burn brightly, creating a luminous haze. Blazing and blinking neon advertising signs reduce the aesthetic atmosphere of an area. Lights running throughout the hours of darkness not only use large amounts of energy by themselves, but the number of industries that can operate throughout the night as a result of artificial light significantly adds to this drain.

The medical procedures made possible by magnification have raised new moral questions. Consider for example, the invitro fertilization of the human ovum. Such a procedure appears highly unnatural to some people. Other processes, such as having children through a surrogate mother, artificial insemination, artificial sex determination of the child at conception and transplants of animal organs into humans, can raise serious moral, psychological and legal questions.

The study of light has given mankind many advantages. It has yielded information about the moon, the planets and the stars as well as the depths of the oceans. How it is used in the future is certainly worthy of study.

Problems

1. A concave mirror has a focal length of 10 cm. If an object is placed halfway between f and $2f$, describe the image.
2. Describe the image in a concave mirror of focal length 10 cm if the object is placed halfway between f and the mirror.
3. Describe the image in a concave mirror with a principal focus of 10 cm when the object is placed at $2f$.
4. If light enters the surface of a liquid at an angle of 30 degrees to the vertical and is refracted at an angle of 45 degrees to the vertical, what is its index of refraction? (You may solve the problem graphically.)
5. What is the index of refraction of a transparent material, if a light ray incident upon its surface at an angle of 45 degrees

to the vertical is refracted at an angle of 60 degrees to the vertical?

6. A convex lens has a focal length of 12 cm. If an object is placed 18 cm from the lens, describe the image.
7. If the object is placed 24 cm from the lens in problem #6, describe the image.
8. If the object is placed 6 cm from the lens in problem #6, describe the image.

Chapter 12

Electrical Charges

Electricity is an integral part of modern life. You usually do not think about how much you depend on electricity until there is a power outage. Then, you not only miss having electric lights but may find it difficult to cook, clean and stay warm (or cool). In personal grooming alone, electricity is used to dry, curl and remove hair, keep one's teeth and gums healthy and clean and press clothing.

As amazing as electricity is, no one has actually seen it. You may think you see electricity when you observe lightning, but that is only the result of an electrical discharge, which ionizes the atmosphere in the same way you ionized tubes of gases in Chapter 11. You cannot smell electricity, only the ozone that results from an electrical discharge such as lightning. However, you have probably felt electricity, or its effect, if you have ever received an electrical shock. In essence, you probably know very little about electricity, other than what it can do for you – or to you!

Static Electricity

During a cold, dry winter evening, you may notice that pulling a t-shirt or sweater over your head in a darkened room produces a crackling sound and flashes of light. Walking across a carpet and reaching for a metal door knob can result in a spark-discharge between your hand and the knob. You most likely have used a plastic comb to pick up small bits of paper after you "charged" it by running it through your hair. This phenomenon is referred to as *static electricity, static charge* or, as in one television commercial, *static cling*. The term *static* was used because, originally, researchers thought the charge did not move. Actually, static electricity is the same electricity you use in your home, except that the number of charges is too small to do appreciable work.

Early researchers developed the foundation for present-day knowledge about electricity through the study of the same static electricity that you experience. In this chapter you will study this phenomenon and other concepts of electricity.

*****

Experiment 12–1. Static Charges

Suspend a vinyl strip and an acetate strip from separate test tube clamps, fastening them with masking tape so the strips can swing freely (Fig. 12.1). Charge the acetate strip by rubbing it with a cotton cloth; charge the vinyl strip with wool cloth. Rub a second acetate strip with the cotton cloth and bring the strip close to each of the suspended strips, being careful not to let them touch. Record your observations. Repeat the process with the wool cloth and vinyl strip. Record your observations.

Recharge the suspended strips. Now rub the acetate strip with the cotton cloth, but this time move the *cloth* just next to the suspended strips. Do you notice any attraction or repulsion? Is it the same effect as with the charged acetate strip? Try it with the wool cloth after rubbing it on the vinyl strip. How do the strips behave compared to the cotton cloth? Formulate a model of charges that is consistent with your observations. How many kinds of charges were you able to identify?

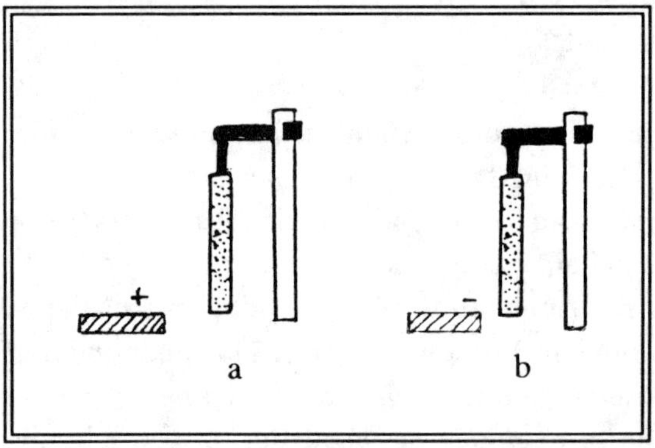

Fig. 12.1. Apparatus for Experiment 12–1.

*****

Conduction

One way to study electrical charges is by using a small styrofoam sphere coated with graphite and suspended on a thin nylon monofilament fishing line, which acts as an insulator. If a hard, rubber rod is charged by rubbing it with a piece of wool and the rod is brought near the sphere, the sphere is attracted to the rod, touches the rod and then bounces away from the rod. If the charged rod is then brought near the sphere, the sphere moves away from the rod. Apparently, the sphere has become charged by the rod. Because contact has been made, the sphere has been charged by *conduction* in that a transfer of charges has taken place. If a second sphere is charged with the rubber rod and brought next to the first charged sphere, the two will repel each other. Because both spheres have the same charge, you have evidence that like charges repel each other.

If a sphere charged by a glass rod that has been rubbed with a silk cloth is brought near a sphere charged by a rubber rod, the two are attracted. Thus, you observe that opposite charges of electricity attract and like charges repel.

In order to distinguish one charge from another, Benjamin Franklin called charges on the rubber rod *negative* and charges on the glass rod *positive*. These were arbitrary distinctions and could just as well have been given distinctions such as *a* and *b* or *x* and *y*. The positive and negative designations are convenient mathematically because $(+1) + (-1) = 0$. When precise measurements are made, one observes that a number of positive charges neutralizes an equal number of negative charges for a net charge of zero. When a body possesses an equal number of positive and negative charges, it is said to be *neutral*.

Using the concept of positive and negative charges, one can examine how the styrofoam sphere became charged by the rubber rod. The sphere, in its neutral state, contains an equal number of positive and negative charges (Fig. 12.2a). As a negatively-charged rod approaches, the positive charges on the sphere are attracted to the side closest to the negative rod and the negative charges are repelled to the opposite side (Fig. 12.2b). When the rod touches the sphere, negative charges move

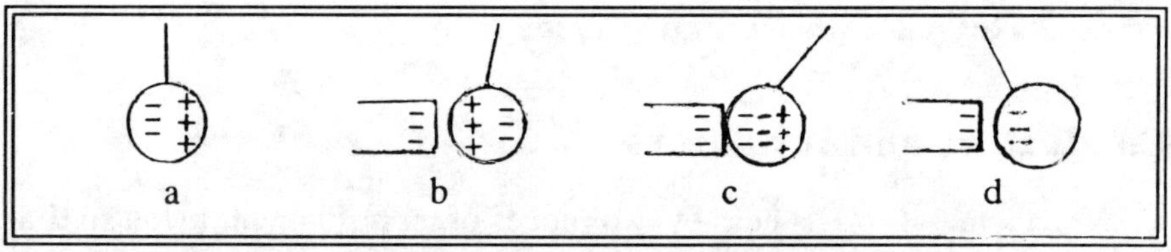

Fig. 12.2. Charging a sphere by conduction.

from the rod to the sphere (Fig. 12.2c); the sphere, which becomes negatively charged, is repelled from the rod (Fig. 12.2d).

If you approach the sphere with the cloth used to charge the rod, you will notice an attraction. Apparently, charges of one kind moved from the cloth to the rod, leaving the cloth oppositely charged.

Induction

Objects can be charged without making contact with the charging rod. A negatively-charged rod can be brought near a neutral, suspended sphere without touching it (Fig. 12.3a). You can provide a path to the earth (ground) by touching the sphere with your finger (You are connected to the earth)(Fig. 12.3b). Negative charges repelled by the negative rod, move to the ground. The sphere is now left with a positive net charge (Fig. 12.3c). When the negatively-charged rod approaches it, the sphere is attracted (Fig. 12.3d).

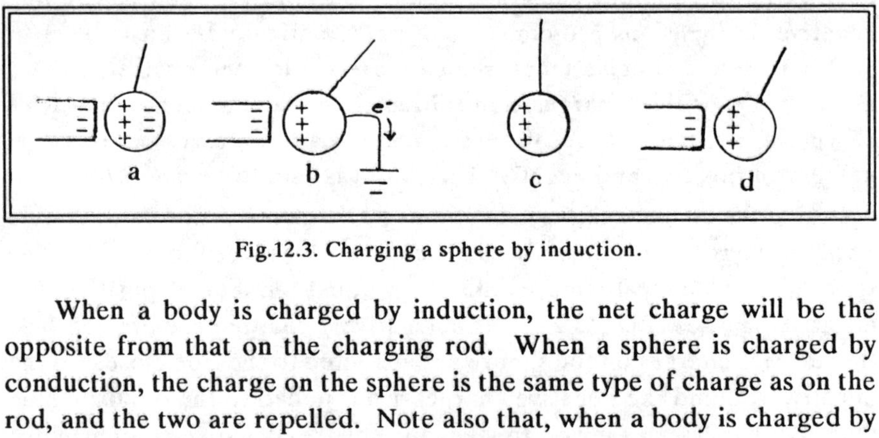

Fig.12.3. Charging a sphere by induction.

When a body is charged by induction, the net charge will be the opposite from that on the charging rod. When a sphere is charged by conduction, the charge on the sphere is the same type of charge as on the rod, and the two are repelled. Note also that, when a body is charged by conduction, an actual transfer of charges takes place, but no exchange of charges takes place when charging by induction.

Conductors and Insulators

When a metal rod is held by two pieces of monofilament nylon so that one end touches a suspended styrofoam, graphite-coated sphere and is touched by a negatively-charged rod, the sphere will take on a charge and bounce away (Fig. 12.4). But, if a glass rod is substituted for the metal rod, the sphere will not be affected. Apparently, some materials conduct electrical charges and others do not. These materials are called *conductors* and *insulators*, respectively.

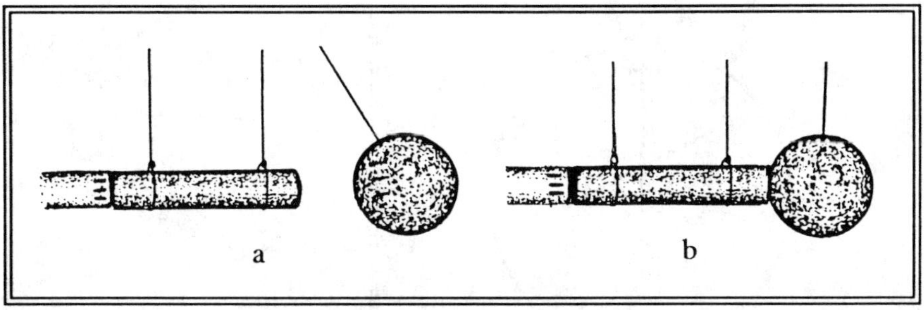

Fig. 12.4. Conductors and insulators.

Conductors have internal charges that are not tightly bound and are relatively easy to move. Thus, charges pass through quite readily. Insulators have very few "free" electrons, are not easily moved by *electrostatic forces* and, therefore, do not conduct charges very well.

*****

Experiment 12-2. Charging by Induction vs. Conduction

Fig. 12.5. A simple electroscope.

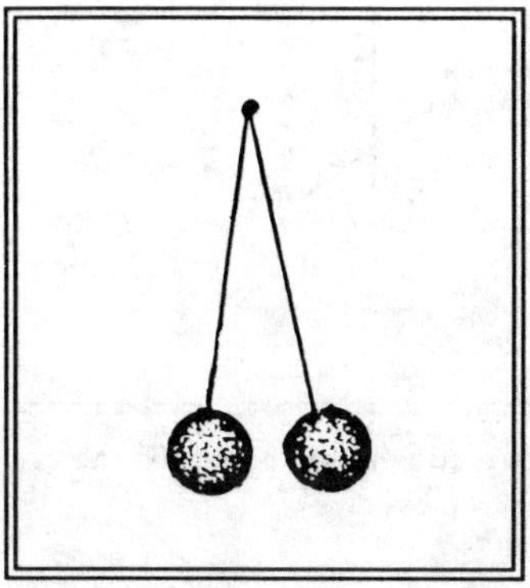

To study the behavior of charges, one needs some type of charge indicator to give evidence of the presence of charges. Such a device, an *electroscope,* consists of two suspended conductors, such as the graphite-coated styrofoam spheres shown in Figure 12.5. However, you will use an electroscope that is easier to handle and has a metal case that shields it from outside electrical fields. It consists of an aluminum-coated section of soda straw suspended so it is free to pivot in a frame. The frame is attached to a metal plate by a machine screw that is insulated from the supporting case by a plastic insulator (Fig. 12.6). The soda straw is mounted slightly off-center so its rest position is vertical. Charges placed on the straw will cause it to be repelled by the like-charged frame.

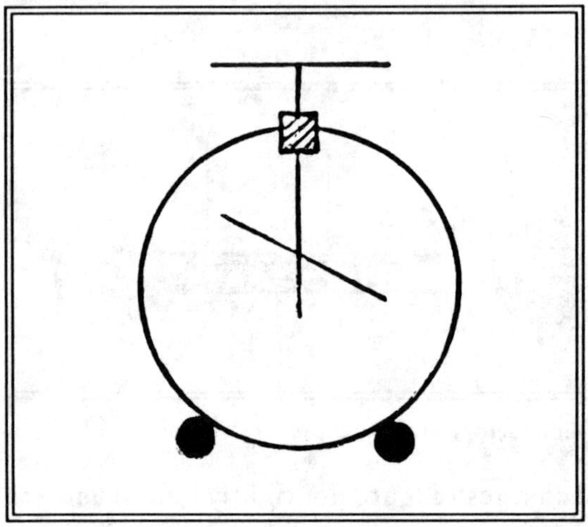

Fig. 12.6. A soda-straw electroscope.

Charge a vinyl strip with a wool cloth and bring it in contact with the *electrode* (disk) of the electroscope (Fig. 12.7). The soda straw should assume a nearly horizontal position (Negative signs have been chosen to indicate a charge). Recharge the vinyl strip and approach the electrode

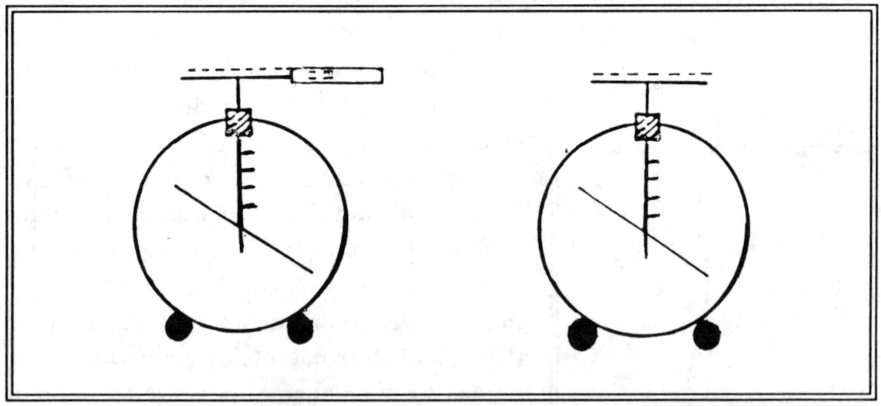

Figure 12.7. Charging an electroscope (a) with a charged vinyl strip and (b) the charged electroscope with the strip removed.

of the electroscope. Do not get so close as to allow a discharge between the strip and electrode. Move the vinyl strip slowly back and forth from the electrode. What is the effect upon the electroscope? Remember, charges were originally transferred to the electroscope by conduction. You are now transferring charges but are inducing a stress upon the charges on the electroscope.

Discharge the electroscope by touching your forefinger to the electrode and your thumb to the frame, providing a path (conductor) for the charges (Fig. 12.8a). Now bring a charged vinyl strip next to the electrode (Do not touch or allow a dischargc) (Fig. 12.8b). With the vinyl strip still near the electrode, momentarily discharge the electroscope by touching your forefinger to the electrode and your thumb to the frame (Fig. 12.8c). Take away the thumb and forefinger; then remove the charged vinyl strip and observe the position of the electroscope straw (Fig. 12.8d). Is the electroscope charged?

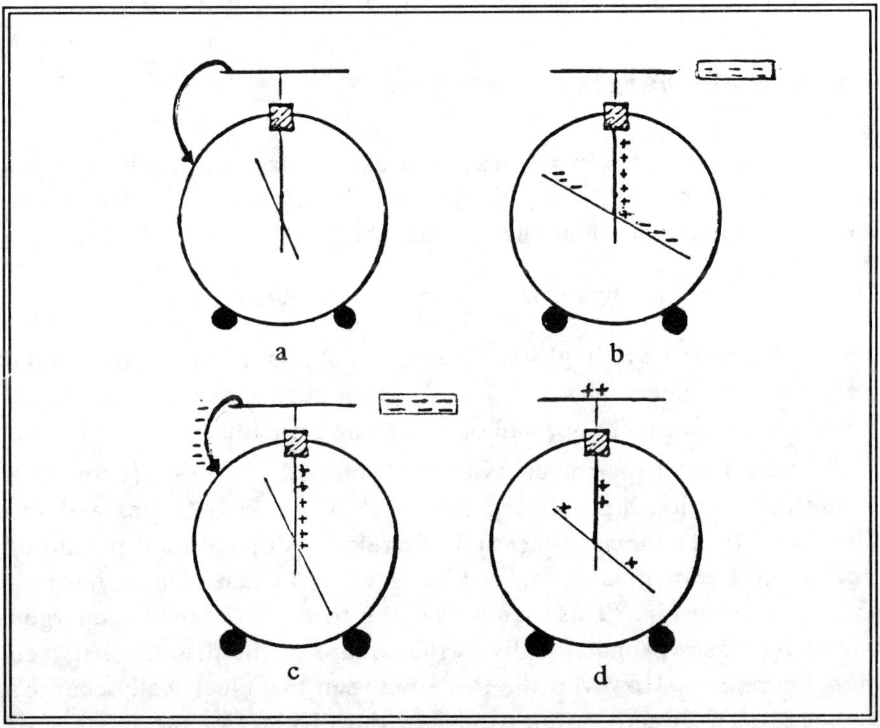

Fig. 12.8.

Approach the electrode with the charged vinyl strip. Observe and record the behavior of the electroscope. How does the charge on the electroscope compare to the charge on the vinyl strip? Try to develop a model to explain the observed phenomenon.

*****

Remember, the electroscope was charged without making contact with the charged vinyl strip. No charges were transferred from the vinyl strip to the electrode. Instead, the electroscope was charged by induction. You have now encountered several instances of electrical charges being influenced across space without making contact. All electrical charges have an area surrounding them that can influence other charges. This area of electrical influence surrounding all electrical charges is known as an *electrostatic field,* because it is *static* (not moving) rather than *dynamic* (moving). Later, you will see that dynamic charges have a magnetic field as well as an electrical field surrounding them.

The Law of Charges

You have observed two characteristics of charges: opposite charges attract each other and like charges repel each other. Therefore an important statement about charges must be:

Like charges repel and opposite charges attract.

This seems to be a simplistic statement, but it is the foundation of the electrical and electronics industries. Even such modern "high tech" machines as computers operate on this basic principle.

Based on his experiments, Charles Coulomb, in 1785, reported a numerical relationship between the force of charged bodies and the distance between them. It turns out the relationship parallels the *law of gravitational attraction,* developed by Newton, in that it is an inverse-square relationship. This means that the force between two charged bodies increases proportionally as the square of the distance between them decreases. However, the force between two electrically-charged bodies is also directly proportional to the product of the number of charges on each body, in the same way the gravitational force between two bodies is proportional to the product of the mass of each body. Coulomb's law expresses the force between charges as:

$$F = k\,\frac{q_1 q_2}{r^2}$$

the symbol *q* represents the number of charges and *r* the distance in meters separating the charges. The constant (*k*) is determined experimentally with units making the equation equal to newtons of force (*F*).

Electrical Current and Potential

Originally, electricity was a curiosity, and research on electrical charges was a part of scientific enterprise that consists of seeking knowledge for the sake of knowledge. As intriguing as electricity was, there was really no "practical" way of utilizing it, although finding applications was a major goal.

The problem with static charges was that not enough of them could be given sufficient potential energy to do a significant amount of work. Perhaps, because the charges were quite small, the question of what kind of work had not even been imagined.

It was not until Alessandro Volta invented the Voltaic pile (battery) in 1800 that enough potential electrical energy could be generated to do practical work, such as the decomposition of metallic oxides to obtain pure metals. As electrical devices were invented, electrical units of measure were developed by empirical experiments for use in their design, manufacture and refinement.

The *volt* is a unit of potential energy per unit of charge at a specific point in a given electrical field. In the MKS system it is designated as 1 *joule of work required to move 1 coulomb of charge from one point to another.* Because, by derivation, one coulomb represents 6.25×10^{18} elementary charges, a volt can be considered as

$$\frac{1\ \textit{volt}}{1\ \textit{coulomb}} = \frac{1\ \textit{joule}}{6.25 \times 10^{18}\ \textit{charges}}$$

$$= 1.6 \times 10^{-19}\ \textit{joules per charge}$$

Thus, just as gravity imparts potential energy (the ability to do work on a body raised above the surface of the earth), *electrical potential* is the ability to do work on individual electrical charges.

The unit of electrical current is the *ampere*. It is by definition one coulomb per second moving past a point. In other words, if you had a "window" in a conductor and you counted 6.25×10^{18} *effective charges* moving by each second of time, you would have witnessed one ampere of current flow. The term *effective charges* is used because an electrical current does not consist of single charges moving from one end of a conductor to the other. The process is more like forcing a marble into a pipe filled with identical marbles and of a diameter slightly smaller than that of the pipe (Fig.12.9). When a marble is forced into one end of the pipe, a marble falls out of the other end. An observer of a window cut in the pipe would witness the movement of one marble past the point. It would not be possible to identify which marble moved by the window.

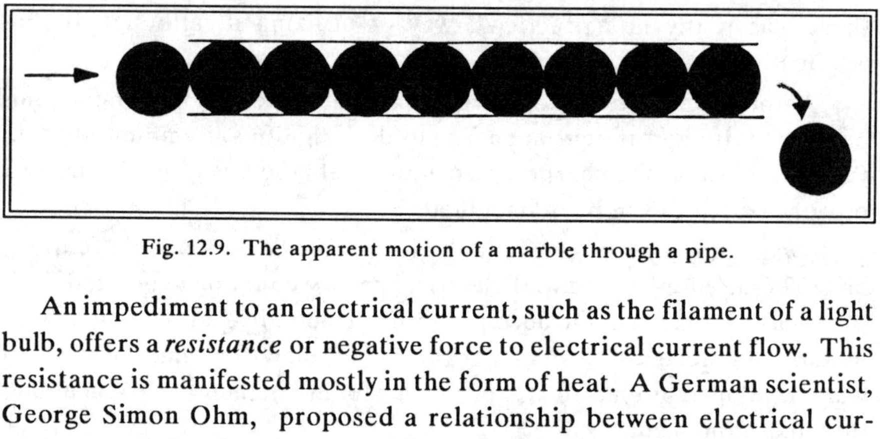

Fig. 12.9. The apparent motion of a marble through a pipe.

An impediment to an electrical current, such as the filament of a light bulb, offers a *resistance* or negative force to electrical current flow. This resistance is manifested mostly in the form of heat. A German scientist, George Simon Ohm, proposed a relationship between electrical current, potential and resistance. *Ohm's law* is expressed as:

$$V = IR$$

where V is the potential in volts and I is the current in amperes. Resistance (R) is measured in *Ohms*, named in honor of Ohm, and is designated by the Greek capital letter *Omega* (Ω).

Figure 12.10 is a drawing of a simple direct current (DC) *series circuit* where the current is unidirectional. A series circuit is one in which the

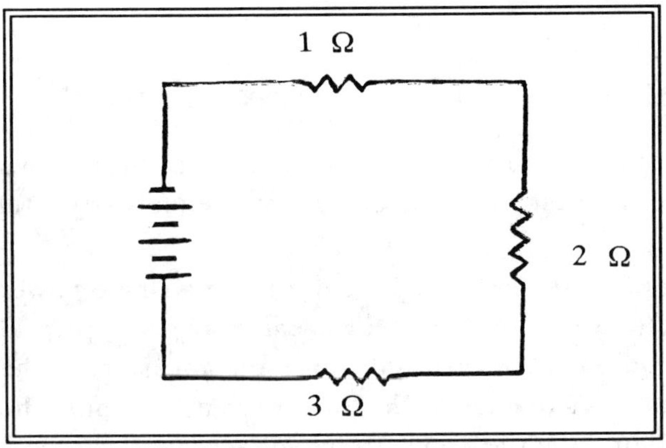

Fig. 12.10. Resistors in series.

current is of the same value in all components of the circuit. The total resistance in the circuit consists of resistors $R_1 = 1\ ohm$, $R_2 = 2\ Ohms$, and $R_3 = 3\ Ohms$. The resistances are additive because all of the current flows through each resistor. Therefore, the total resistance according to Ohm's law is

$$R_t = R_1 + R_2 + R_3 + ...R_n$$

and in this example

$$R_t = 1\,\Omega + 2\,\Omega + 3\,\Omega = 6\,\Omega$$

The total current in the circuit can be determined from Ohm's Law where $V_t = I_tR_t$, and can be expressed as:

$$I_t = \frac{V_t}{R_t}$$

and

$$I_t = \frac{6\ V}{6\ Ohms} = 1\ amp$$

Therefore the potential across each resistor is:

$$V_1 = I_tR_1 = 1\ amp \times 1\ Ohm = 1\ volt$$
$$V_2 = I_tR_2 = 1\ amp \times 2\ Ohms = 2\ volts$$
$$V_3 = I_tR_3 = 1\ amp \times 3\ Ohms = 3\ volts$$

and

$$V_t = V_1 + V_2 + V_3 = 1 \text{ volt} + 2 \text{ volts} + 3 \text{ volts} = 6 \text{ volts}$$

which is the potential of the battery (6 *volts*) applied to the circuit. In a series circuit, the sum of voltage across each resistor equals the total voltage at the source.

This principle is important when you consider using a single power source such as a 12-volt battery to supply different voltages to several pieces of electrical equipment located at some distances from the battery. You could run wires to each of the locations and compute the resistances required to provide the appropriate voltages at each location. For example, power lines are really mechanisms used to bring an electrical potential into your home. They are used instead of bringing the electrical generator to your home or your going to the powerhouse to plug in your electric toothbrush.

The relationship between potential, current and resistance is different in a *parallel circuit* (Fig. 12.11) than in a series circuit. Here the total current does not flow through each resistor, but is divided between the three proportional to their respective resistances. However, the potential (voltage) is the same across all three resistors because the terminals of the battery are connected directly across them.

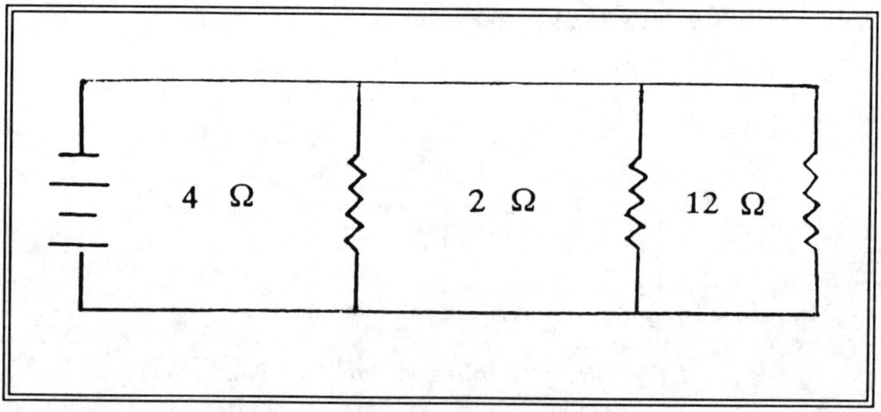

Figure 12.11. Resistors in parallel.

Therefore, according to Ohm's law, the current in each resistor is:

$$I_1 = \frac{V_t}{R_1} = \frac{6\,V}{4\,\Omega} = 1.5\ amps$$

$$I_2 = \frac{V_t}{R_2} = \frac{6\,V}{2\,\Omega} = 3.0\ amps$$

$$I_3 = \frac{V_t}{R_2} = \frac{6\,V}{12\,\Omega} = 0.5\ amp$$

Because the current flow in one component of a circuit combines with the current flow of all the other components to form the total current, the total current in this example becomes:

$$I_t = 1.5\ amps + 3.0\ amps + 0.5\ amps = 5.0\ amps$$

The resistance provides three paths for the current to flow, making the total resistance less than that due to either individual resistor. Measuring the total resistance with an ohmmeter gives a total resistance of 1 1/5 *Ohms*. Mathematically the total resistance in a parallel circuit is computed by the relationship:

$$\frac{1}{R_t} = \frac{1}{R_1} + \frac{1}{R_2} + \frac{1}{R_3} + \ldots \frac{1}{R_n}$$

Substituting the resistence values above yields:

$$\frac{1}{R_t} = \frac{1}{4\,\Omega} + \frac{1}{2\,\Omega} = \frac{1}{12\,\Omega} = \frac{3+6+1}{12\,\Omega} = \frac{10}{12\,\Omega} = \frac{5}{6}$$

Because $R_t = 6/5\ \Omega$, the total current flow is $6\ V \div 6/5$ Ohms = 5.0 *amps*, exactly what it came out in calculating individual current flows and adding them.

Electrical Power

Electrical power, the time-rate of doing work, is determined by the relationship $P = VI$. Recalling the derivation of the volt and the ampere, leads to:

$$P = \frac{joule}{q} \times \frac{q}{sec} = 1\ joule/sec = 1\ watt$$

Returning again to Figure 12.10, the total power is:

$$P = VI = 6\ V \times 1\ amp = 6\ watts$$

The power in each resistor is:

$$P_1 = V_T I_1 = 1\ V \times 1\ A = 1\ watt$$
$$P_2 = V_T I_2 = 2\ V \times 1\ A = 2\ watts$$
$$P_3 = V_T I_3 = 3\ V \times 1\ A = 3\ watts$$
$$P_T = 1\ watt + 2\ watts + 3\ watts = 6\ watts$$

The power dissipation in a series circuit is additive. This is also true in parallel circuits, where the current through each component is multiplied by the voltage across each respective component and added to equal the total power. An alternative approach is to multiply the total potential (volts) across the circuit by the total current (amps) flowing through the circuit. In some cases, it is desirable to know the power dissipated in each component.

*****

Experiment 12-3. Series Light Bulbs

For this experiment you will use a battery, three bulbs, three sockets and four wire leads. Connect the bulbs to the battery according to the schematic diagram in Figure 12.12. Be sure to leave one battery terminal disconnected and touch it only momentarily to light the bulbs. Dry cells are not designed to operate for more than short periods of time. Note that you have four dry cells in series, with each cell of 1.5 *V*, to equal a total potential of 6 *V*. The short bar of a cell in the schematic diagram is the negative terminal and the longer bar is the positive terminal of a cell. A battery is technically two or more cells connected to aid each other. Each

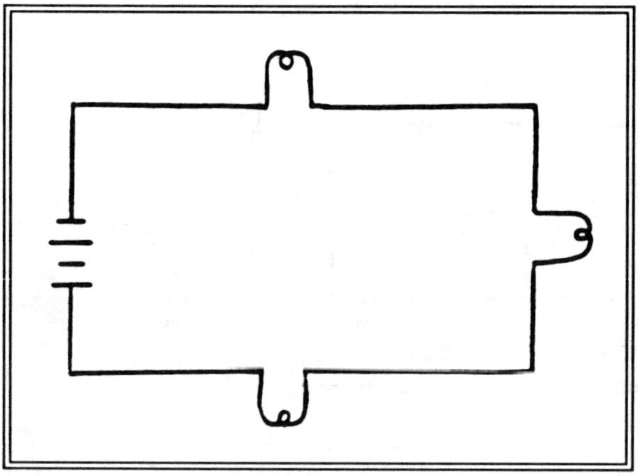

Fig. 12.12. Three light bulbs in series.

light bulb is represented by a wire loop indicating a filament.

With the circuit connected, unscrew one bulb until it is no longer lit. What happens to the other bulbs? Tighten the bulb back into its socket and unscrew a second bulb and observe the other two. Screw bulb number two back in and unscrew the third bulb. What can you say about a series circuit? Screw the third bulb back in and observe the intensity of three bulbs in series. Now connect only one bulb to the battery momentarily and contrast its intensity to that of three bulbs in series. Contrast the intensity of two bulbs in series to three bulbs in series. What happens to the intensity when bulbs of equal size are added to a series circuit?

*****

Experiment 12–4. Parallel Light Bulbs

Connect the circuit as shown in the schematic diagram of Figure 12.13. Momentarily connect the circuit and compare the intensity of the bulbs to the intensity of the three bulbs connected in series in the previous experiment. Now, with the circuit connected, unscrew one of the bulbs and carefully observe the other two bulbs. Unscrew a second bulb and observe the effect on the remaining bulb. What general rule can you make about removing bulbs and intensity in parallel circuits?

*****

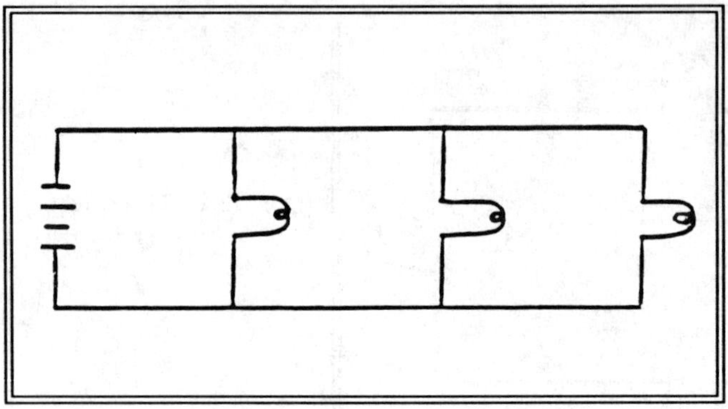

Fig. 12.13. Three bulbs in parallel.

Short Circuits/Open Circuits

Short circuits, or simply "shorts," are the result of a conductor inadvertently bypassing the components in a circuit, thus causing the interruption of electrical service. If a copper wire accidentally comes across the leads connected to one of the light bulbs in Figure 12.13 (or if the insulation is removed from a portion of the leads and the bare spots come in contact with each other), the current will bypass the lamps and they will no longer be lit. Because the resistance is essentially zero, the current will become very high and the shorted circuit will become very hot. In the case of a house circuit, or even in the circuitry of an automobile, the current can become hot enough to cause a fire.

The possibility of a short circuit is the reason most circuits have a fuse or automatic circuit breaker in them. A fuse is a device that melts when the current reaches a critical level, thereby opening the circuit just as a switch opens a circuit so that current flow ceases. Automatic circuit breakers do this electromagnetically.

Sometimes an electric appliance, such as a light bulb, will operate sporadically and then cease to function. Many people consider this a short circuit, when, in fact, it is an *open circuit*. In other words, the complete circuit is broken. The reason a bulb blinks prior to going out is that contact is made occasionally due to vibration, such as with a flickering automobile tail lamp. When the circuit is permanently broken, such as with a burned-out filament, the current ceases to flow. Keep in mind that short circuits result in heat being given off and can be dangerous; whereas open circuits simply interrupt electrical service. If arcing occurs with a loose connection, however, ignition of flammable materials can result.

Capacitors

Early experimenters developed "jars" that could store electrical charges. These *Leyden jars* (named after the town in the Netherlands where they were first developed) are glass jars with metal coatings on the inside and outside walls. The coatings are insulated from each other by the walls of the glass jar. A conductor connected to the inside of the jar passed through an insulator in the mouth of the jar (Fig. 12.14) to provide an accessible terminal to the inside metal coating.

Fig. 12.14. Leyden jar.

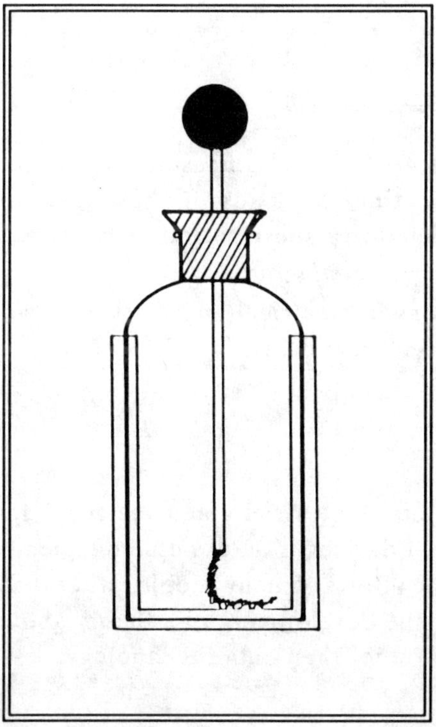

Charges continuously placed on the center electrode of the jar build up considerable potential between the inside and outside metallic coatings on the jar. When one end of a conductor is connected to the outside metallic coating and its opposite end brought very near to the center electrode, a large discharge would give evidence of a huge stored charge.

The Leyden jar has been replaced by the modern *capacitor.* A capacitor consists of two conductors separated by a *dielectric* (insulator). The symbol for a capacitor is given in Figure 12.15. Connecting a battery to the capacitor causes each of its plates to assume the polarity of the battery. The potential across the capacitor is initially zero. Through induction, the negative potential on one plate attracts the positive charges on the opposite plate which is connected to the positive terminal of the battery. Positive charges migrate quickly to the positive terminal, registering a current flow in the circuit, even though no charges flow between the plates. Soon the positive plate of the capacitor is saturated with positive charges and the negative plate is saturated with negative charges. The potential between the plates of the capacitor reaches the potential of the applied potential of the battery, resulting in a total curtailment of the current flow.

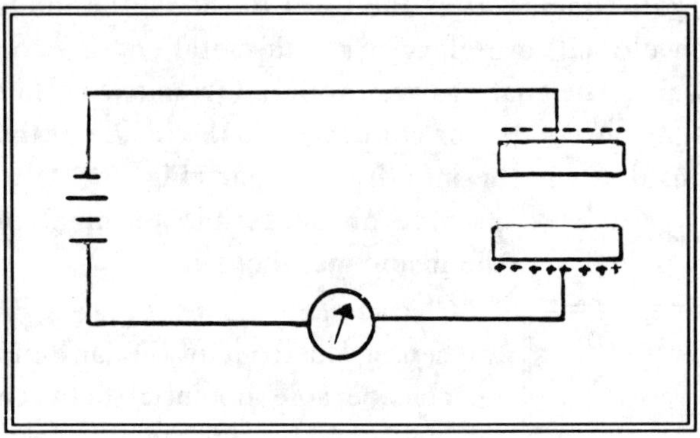

Fig. 12.15. Charging a capacitor with a battery.

If the battery is disconnected, a potential can be measured across the terminals of the capacitor. Connecting the leads of the capacitor together registers a high current flow for a short period of time on a current meter connected in series.

Capacitors have many applications in electrical and electronic circuitry, including voltage regulation and time delay.

Science/Technology/Society

The development of knowledge about electricity and the resultant invention of electrical machinery and devices is an example of science and technology advancing through the efforts of many people rather than the efforts of one person. Much of the development of a theory about electricity came out of pure science rather than out of technology.

One is hard-pressed to think of a modern machine that does not depend upon electrical theory. Most internal combustion engines, with the exception of diesel engines, depend upon the electrical ignition of combustible gases. Steam turbines do not use electricity for operation, but the servomechanisms used in their regulation require electricity.

Electrical and electronics industries, such as those producing the silicon chip used in microprocessors, seem to spring up and mushroom over night. All industries require space as well as energy and usually create some type of pollution. Judicious consideration should be exercised in locating industrial plants so that such factors as prevailing atmospheric conditions, aquifers, water run-off conditions and the presence of other polluters are not ignored. Profit should not be the only deciding factor.

Present-day technology seems able to produce any gadget mankind could desire, from the most trivial to those necessary for survival in a modern society. Being scientifically literate allows individuals to make wise consumer decisions. Purchases should be made more on the basis of "needs" than "wants." Some questions that can serve as guidelines in making purchases are: Is the item recyclable? Is it designed with "planned obsolescence within an unreasonable replacement time"? Will a hand-powered device, such as a manual can opener, do the job just as well as the electric model?

The waste of electrical energy by individual consumers is especially acute. It seems that people increasingly believe that electrical energy is limitless, and some people consistently demonstrate their ignorance of what individual acts imply. You can routinely walk through buildings and see electric lights burning when no one is in the room. In homes, one not only sees electric lights burning in rooms with no occupants, but may also encounter operating television sets, radios or audio players.

Climate control of living environments is also problematic. The tendency is to overheat in the winter and over-air-condition in the summer. Most people would probably be able to adapt somewhat to temperature extremes if their bodies were given some time to adjust. In fact, one researcher has suggested that one can adapt to the summer heat better if one takes warm showers to allow the body to adjust to the ambient temperature in the summer. The simple acts of turning off appliances and lights when they are not in use, using electric fans and the cooling breezes in summer and wearing warmer clothing indoors during the winter can reduce the waste of electrical energy.

People also need to plan the time they use electricity. Electricity cannot be stored easily during low-demand periods to compensate for use during peak-demand periods. To compensate for this factor, power plants are built much larger than would normally be required in order to compensate for increased energy demands during peak-demand periods. Cooking, washing and drying clothing, and temperature control could be staggered, instead of everyone attempting to do these things at the same time.

Electricity has been a tremendous boon to mankind. The benefits and conveniences derived from the applications of electricity seem almost limitless. However, electricity should not be treated as a limitless commodity. Each watt of electricity generated extracts a tremendous price in terms of natural resources and the effect upon the environment. The natural resources and land surface area of the planet are finite. As the population of the world increases, the per capita amount of these resources becomes less. As more pollution enters the atmosphere, the amount of available clean air for breathing becomes less.

Problems

1. Describe how you would charge an electroscope inductively with a negatively-charged rod such that the electroscope is left with a positive charge.
2. 10-Ohm, 15-Ohm and 25-Ohm resistors are connected in series to a 100-V battery. What is the total current in the circuit? What is the current in the 15-Ohm resistor? What is the voltage across each resistor?
3. 5-Ohm, 10-Ohm and 15-Ohm resistors are connected in series to a 60-V battery. What is the total resistance in the circuit? What is the total current-flow in the circuit? What is the voltage across each resistor?

4. What is the power dissipated across each resistor in Problem #3? What is the total power dissipated in the circuit in Problem #3?

5. A 5-Ohm, a 10-Ohm and a 15-Ohm resistor are connected in parallel to a 50-V battery. What is the total resistance in the circuit? What is the total current flow in the circuit? What is the current flow in (a) the 5-Ohm resistor, (b) the 10-Ohm resistor and (c) the 15-Ohm resistor? What is the voltage across the 15-Ohm resistor? What is the power dissipated by the 10-Ohm resistor?

6. Define electrical potential. What is the unit of electrical potential?

7. Define electrical current. What is the unit of electrical current?

8. Define electrical resistance. In what unit is electrical resistance measured?

9. What is a capacitor? (a) What happens when the terminals of a battery are connected across the terminals of a capacitor? (b) What happens when the terminals of a charged capacitor are brought near each other?

10. How would you connect a voltmeter to a resistor in a circuit to measure the potential across it?

11. How would you connect a current meter (ammeter) in a circuit to measure the current flowing through a light bulb?

12. Light bulbs are glowing in an electrical circuit. The filament in one bulb burns out and the other bulbs go out. Describe what type of circuit this is.

13. When you open one portion of a circuit consisting of three light bulbs, two bulbs go out but one bulb remains burning. Describe the circuit

14. Describe an *open circuit*.

15. Describe a *short circuit*.

16. Two spheres, one with a positive charge of 10 coulombs and the other with a negative charge of 50 coulombs, are located 10 mm from each other. (a) What is the force in Newtons between them? (b) Is the force attractive or repulsive?

Chapter 13

Magnetism and Electricity

As early as 500 B.C., it was observed that stones of black iron ore with a bluish tint were attracted toward iron. This iron ore, originally found near the city of Magnesia in western Turkey, became known as *magnetite*. Around 1200 A.D., Europeans noted that chunks of magnetite suspended on a string would align themselves in a north-south direction. When they were rotated 180° from the rest position, they would return to the original north-south position with the same end pointing north. The north-seeking end was marked *north*. Because magnetite stones always aligned in the direction of the earth's magnetic influence, the north-seeking end became known as the *north pole* and the south-seeking end as the *south pole*. These stones were used in navigation to identify north and became known as *lead stones* or *lodestones*. As such, they were the forerunners of the modern compass. Objects which attracted iron became known as *magnets* and the property of attraction as *magnetism*.

*****

Experiment 13–1. The Law of Magnets

Suspend a bar magnet from a piece of string (Fig. 13.1) and allow it to rotate back and forth until it comes to rest. Record observations as to direction of alignment and the direction each pole is pointing. Pull one

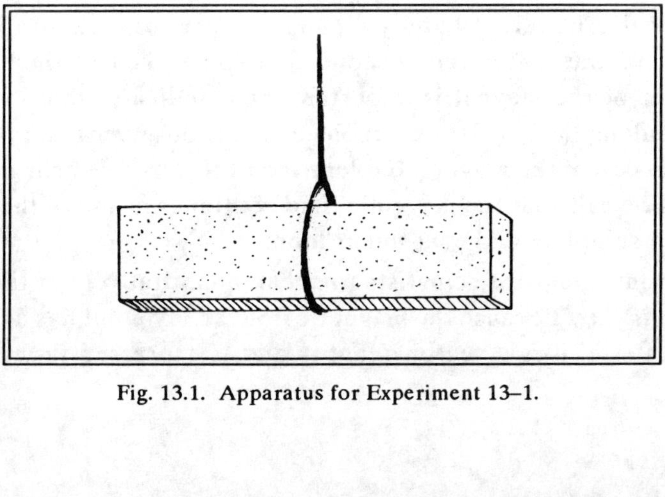

Fig. 13.1. Apparatus for Experiment 13–1.

end of the magnet aside and repeat the process. What do you notice about the position of the magnet each time?

Allow the magnet to come to rest on its own. Approach the north end of the suspended magnet with the north pole of another magnet. What do you observe? Now approach the north end of the suspended magnet with the south pole of the other magnet. What do you observe? Repeat the process with the south pole of the suspended magnet. What do you observe? State a general law for magnets.

*****

The Magnetic Field

The area of magnetic influence around a magnet is called its *magnetic field*. When powdered iron or iron filings are placed around a bar magnet, the iron particles line up in a pattern parallel to the stones but curving in and concentrating at the poles. The poles of a magnet not only line up with the earth's magnetic poles but are also the areas on a magnet where the magnetic force concentrates.

Imaginary lines of force formed by the positions of the iron filings in the magnetic field are called *magnetic flux*. You can construct the magnetic flux by drawing a line tangent to each individual iron filing at its point of rest in the field of a magnet. The direction of the flux can be determined by placing an arrowhead on the end of the tangent line closest to the south pole. The magnetic field, by convention, is considered to travel from the north pole to the south pole of a magnet. A quasi-circular pattern emerges around the bar magnet. It appears to radiate from the center and fans out to the poles. Noting the pattern at the center of the magnet is important, as it indicates that magnetism exists all along the bar. However, one can only observe what takes place on the outside of the magnet. Exploration of the inside of a magnet with neutrons reveals that the magnetic field continues inside of the magnet, forming a complete and continuous loop.

It is convenient for scientists to speak quantitatively of the *density of magnetic flux*. Because the magnetic field at any point has a direction as well as size, it is a vector quantity and is expressed in webers per square meter, an arbitrary unit usually equal to one flux line per square meter. The relationship for the magnetic flux density (B) is expressed mathematically as:

$$B = \frac{\Phi}{A}$$

where Φ represents the lines of flux perpendicular to the magnetic field at a given point and A is the cross-sectional area at that point. A weber per square meter is also known as a *tesla.*

Charles Coulomb was the first to study the strength of the magnetic force between two poles of a magnet. He found this force was directly related to the product of the strength of the poles and inversely proportional to the square of the distance between them. This relationship:

$$F = \mathrm{k}\frac{M_1M_2}{\mathrm{d}^2}$$

parallels his law for static charges discussed in Chapter 12.

Permanent and Temporary Magnets

If a piece of hard steel is rubbed from one end of a magnet to the other, it becomes magnetized and remains so when removed from the magnet. The steel has become a *permanent magnet.* A soft-iron rod placed against the pole of a permanent magnet also becomes a magnet and will pick up iron filings. However, the soft iron rod is a *temporary magnet*, because as soon as the permanent magnet is removed, it loses most of its magnetism. Only a very slight amount of "residual" magnetism remains for a short time. Due to its ability to assume magnetism and then lose most of it on demand, soft iron has many industrial applications, i.e., transformer cores, electromagnets, electric generators and electric motors.

Permanent magnets can also be formed by placing hard steel in a strong magnetic field and carefully moving it out of the field in a direction parallel to the poles. Striking a hard steel rod aligned parallel to a magnetic field will also cause it to become a permanent magnet. Conversely, striking a permanent magnet not aligned parallel to a magnetic field can demagnetize it. Heating a permanent magnet can destroy its magnetism. Permanent magnets should not be struck harshly or heated and should be stored in pairs with their opposite poles attracted to each other or, as in the case of U-shaped magnets, with a piece of soft iron, known as a *keeper*, attached from pole to pole.

Magnetic Substances

Iron, steel, nickel and cobalt are some of the relatively few substances strongly attracted by magnetic fields, with iron and steel being affected more than the other two. These members of the iron family that can maintain magnetism outside the influence of a magnetic field are said to be *ferromagnetic*, from the Latin word *ferrum* (iron). Substances that do not exhibit ferromagnetic properties are designated *non-magnetic*.

Very powerful modern magnets are constructed by combining metals or metallic oxides. Permalloy is an alloy containing from 30 to 90 percent nickel. AlNiCo is a combination of aluminum, nickel, cobalt and iron. Such magnetic alloys can support weights at least 1000 times their own weight. Iron oxides are combined with other basic metal oxides (i.e.,cobalt, copper, magnesium, manganese, nickel and sodium) to form *ferrite* compounds, which become powerful magnets when fired under heat and pressure in a strong magnetic field. Ferrites can be formed into desired shapes and exhibit very high electrical resistances that allow for unique research and industrial applications.

The Magnetic Domain Theory

In Chapter 10 matter was described as basically electrical in nature, with a positively-charged nucleus surrounded by a negative electron cloud. In 1820 Hans Christian Oersted, a Danish physicist, demonstrated that an electric current flowing through a wire deflected a magnetic compass. He suggested there was a magnetic field associated with a moving charge. About a decade later, Michael Faraday in England and Joseph Henry in the United States discovered that a magnetic flux moving through an electrical conductor generated an electrical potential in the conductor. Thus, electricity and magnetism were linked together.

The present theory of magnetism links the magnetic properties of matter to the motion of electrons in the atoms. Two kinds of electron movement lead to magnetic properties of an atom. One is the revolution about the nucleus, which leads to weak diamagnetism, and the second is electron rotation or *spin*. Electrons tend to form pairs with opposite spins that cancel the magnetic effects of each. Elements whose atoms have an imbalance between electron orbits and spins due to electron configuration tend to be *paramagnetic*.

Ferromagentic substances have unpaired electrons in their outer orbits, all with like spins. It is believed that electrons are responsible for the ferromagnetic properties of the atoms. However, this would suggest that all pieces of iron are permanent magnets, which is definitely not the case. Although each atom acts as if it were a permanent magnet, not all atoms are oriented in the same direction. A close inspection of iron stained with iron oxide indicates a crystal structure. Each crystal of iron consists of groups of atoms with random magnetic orientations. These groups of atoms in the crystalline structure of iron, known as *magnetic domains*, cancel out their magnetic properties by their random alignment. If the iron material is placed in a magnetic field, the domains align themselves parallel to the field with the magnetic fields of the domains reinforcing each other. The domains merely rotate about a fixed point during alignment and do not change their location. If the domains

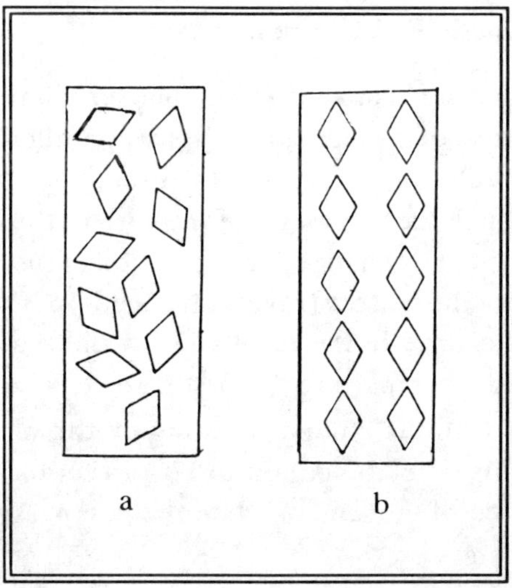

Fig. 13.2. Alignment of magnetic domains: (a) unmagnetized iron bar and (b) magnetized iron bar.

remain aligned after the magnetic field is removed, the magnet becomes permanent, but if they lose their alignment, the net magnetic effect is lost and the material becomes demagnetized.

In Figure 13.2 the diamond shapes represent magnetic domains. In reality these domains are crystals whose shape depends upon the type of material. Notice that in Figure 13.2a the domains are randomly oriented about fixed positions. Figure 13.2b represents the domains placed in a magnetic field. Note that they have rotated such that their north poles are aligned in one direction. The magnetic field of each domain reinforces the field of the other, creating a magnetized piece of material, or a magnet. If the magnet is sliced into small segments, each segment is a magnet with a north pole and a south pole.

Magnetic Fields of Direct Currents

The relationship between magnetism and electricity was not realized for centuries, as a relationship between static charges and static magnetic fields could not be made. In 1820, however, Hans Christian Oersted demonstrated the magnetic effect of moving charges in a wire. You can experience Oersted's demonstration by doing Experiment 13–2.

*****

Experiment 13–2. The Magnetic Field Around a Wire

Connect one end of a 20-cm length of an insulated wire to one terminal of a battery. Lay the wire along the face of a magnetic compass, parallel to the alignment of the compass needle, in a north-south direction as shown in Figure 13.3. Momentarily touch the other end of the wire to the opposite terminal of the battery. Observe any change in the position of the compass needle. Reverse the polarity of the battery (reverse the leads) and repeat the process. Did you notice a change in the results? How did the direction in the movement of the needle compare to the first try?

Repeat the experiment, this time with the compass resting on top of the wire. Was the needle deflected this time? If so, how did its direction of movement compare to the direction of movement when the compass was under the wire?

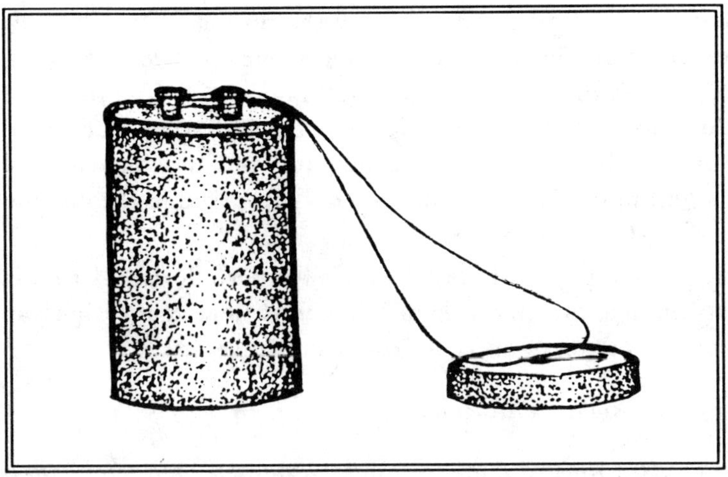

Fig. 13.3. Relationship of an electric current to a magnetic field.

If the needle was deflected when the current flowed through the wire, what must have been present around the wire? If a magnetic field is present around a wire, what must be true of the electric charges inside the wire? If electric charges are moving in a wire, what is present around the wire?

*****

In doing Experiment 13–2, you have demonstrated that an electrical current flowing through a wire has a magnetic field that encircles the wire. The question becomes: Is the magnetic field due to the charges flowing through the wire or is it due to some inherent characteristic of the wire? To answer this question, you can run an electrical current through an electrical wire as you did in Experiment 13–2 and measure the magnetic field. You can repeat the experiment by running exactly the same current through a tube of electrolyte (a solution that conducts an electrical current) and again measuring the magnetic field. It turns out that for identical currents, the magnetic fields are the same. Any moving charge has an associated magnetic field.

The shape of the magnetic field around moving charges can be demonstrated easily in the laboratory. A copper wire is run vertically through a horizontal plane of insulating material such as glass or cardboard. When the battery is connected to the conductor, iron filings, which have been sprinkled evenly on the insulating material, line up in concentric circles around the wire. Thus, one observes that the magnetic field encircles the conductor at right angels to the motion of charges. The direction of the magnetic field can be charted by placing a compass at various points around the conductor and drawing arrows on the surface of the insulator. If the current is reversed in the conductor, the magnetic field will also reverse.

The French physicist André Marie Ampère devised a rule of thumb for determining the direction of a magnetic field in a straight conductor when the direction of current flow is known:

Ampère's Right Hand Rule:

> *Place the fingers of the right hand around a conductor with the thumb pointing in the direction of the electrical current flow and the fingers will point in the direction of the magnetic field encircling the wire.*

The right hand rule assumes current flow as positive to negative. If electron flow is considered, the same rule applies, using the left hand.

The Solenoid

The magnetic field surrounding a straight conductor is too weak to use for anything but analysis and the calibration of electrical apparatus.

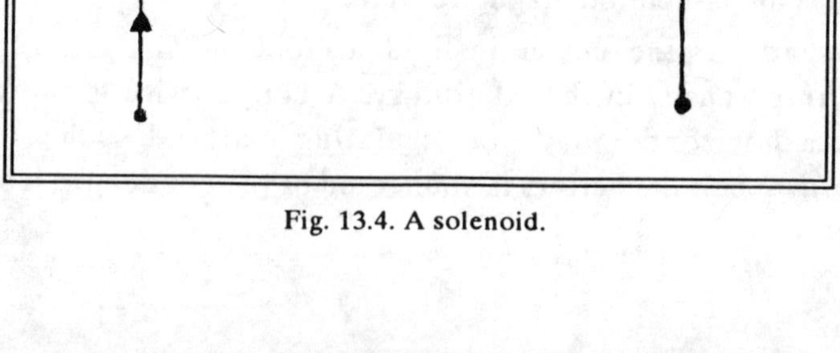

Fig. 13.4. A solenoid.

To increase the magnetic effect of an electrical current, the conductor can be made more compact by forming a loop. Running a current through a loop of wire does, indeed, set up a magnetic field at right angles to the plane of the loop with a north pole and a south pole.

By winding a series of loops around a cardboard core, such as those found at the center of a roll of paper towels, you can produce a strong magnet known as a solenoid. In fact, the wire can be wound around the tube several layers deep, as long as the windings continue in the same direction.

The magnetic field in each loop of a solenoid has the same current flowing through each loop in the same direction. Thus, the magnetic field of all of the loops is oriented in the same direction and reinforces each other to form a powerful magnet. Ampère's right hand rule for current flow can be adapted to become Ampère's right hand rule for a solenoid:

Ampère's Right Hand Rule for a Solenoid:

> *Grasp the solenoid with fingers of the right hand, encircling the solenoid in the direction of current flow, and the thumb will point in the direction of the north pole of the magnetic field.*

The strength of the magnetic field of a solenoid is directly related to the current flow and the number of windings. Because flux is dependent upon cross-sectional areas of the field, the magnetic field is inversely proportional to the radius of the windings.

The Electromagnet

Solenoids, as described in the previous section, are not very useful other than for research purposes. However, the solenoids with movable iron cores are quite useful in applications in servo-mechanisms.The permeability of the air core is not conducive to a strong flux. In Experiment 13–3 you will construct a solenoid with an iron core, called an *electromagnet.*

*****

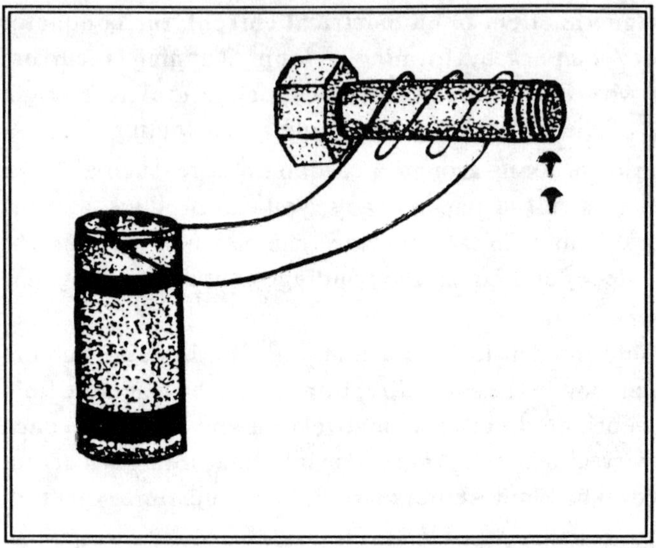

Fig. 13.5. An electromagnet.

Experiment 13–3. The Electromagnet

Wrap about 20 turns of insulated wire around a bolt as shown in Figure 13.5. It is all right to form more than one layer, just be sure you wrap all turns in the same direction and that the turns are wrapped closely together and as tightly as possible.

Connect the stripped ends of the wire to a battery and see how many small tacks you can pick up with the end of the bolt. Disconnect one end of the wire from the battery. What happens to the tacks? See how many tacks you can pick up by increasing the number of turns on the bolt to 30 turns and then 40 turns.

Keep the 30 or 40 turns of wire on the bolt and connect two batteries in series to double the voltage. How does this affect the number of tacks the magnet will pick up? Connect three batteries and then four to determine the number of tacks the electromagnet will pick up. How does the increase in voltage affect the strength of the electromagnet? What generalizations can you make from this experiment? What are some of the characteristics of electromagnets?

*****

An electromagnet consists of a *coil*, an iron *core* and an electrical *current*. In Experiment 13–3 you observed that the strength of an electromagnet is directly proportional to the electrical current and the number of turnings. It is also dependent upon the permeability of the iron core.

Electromagnets have the advantage of being temporary magnets. When the current is stopped, most of the magnetism ceases. Heavy loads of iron and steel can be loaded and unloaded by electromagnets on the end of a crane. When the desired location is reached, the current is interrupted and the iron is dropped. Electromagnets also have important uses in electric motors and electrical generators. Special electromagnets (*solenoids*) with movable iron cores have many applications in servo-mechanisms where mechanical movement is done electrically at a remote location. Electromagnets are an integral part of electrical measuring devices such as the galvanometer.

The Galvanometer

You will recall that an electric current moving through a wire caused a deflection on a magnetic compass. The stronger the current in the wire, the greater the deflection of the needle.

The *galvanometer* is a very sensitive device that uses the same principles to measure very weak electrical currents. To give the galvanometer a high degree of sensitivity to small currents, the poles of a strong stationary *field magnet* surround a small movable coil mounted on a horizontal axis with a pointer attached to it. A small current flowing through the movable coil creates a weak magnetic field that is repelled by the powerful field magnet. This causes the coil and its attached pointer to rotate. The electrical units can be measured as the pointer passes over a scale. The coil rotates instead of the field magnet, and, because of its low inertia, is more responsive to extremely small currents. The pointer on the galvanometer comes to rest at the center of the scale and can swing toward either end, depending on the polarity of the applied current.

Galvanometers are used as very sensitive devices for calibration and laboratory research. They also form the basis for the *ammeter*, a device for measuring currents in electrical circuits. An ammeter is placed in series with the circuit under measure, causing all of the current to pass through a low-resistance resistor. The resistor must be of a size so that only a small amount of current actually passes to the galvanometer. The galvanometer's pointer is at rest at the zero mark on the left of the scale, which is calibrated in amps, milliamps or microamps to meet the needs of the circuit under measure.

A galvanometer is also the central device in a *voltmeter*. The voltmeter measures the potential across a component in a circuit and, thus, has a high internal resistance that allows only a minute amount of current to be drawn from the circuit and directed to the galvanometer.

The Magnetic Field of the Earth

The earth itself behaves as a magnet. When you suspended the bar magnet on a string, it oscillated in a horizontal plane for a few moments and came to rest in a north-south position. You probably noticed that the end of the magnet marked *N* was pointing in a northerly direction. This end is called the *north pole* of the magnet because it points to the north magnetic pole of the earth. Keeping in mind that opposite poles attract, it becomes apparent that this end of the magnet is of opposite polarity from the north magnetic pole of the earth. Thus, the north pole of the magnet is really a *north-seeking* pole. Because of magnetic anomalies in the earth itself, the north magnetic pole of the earth does not always coincide with the rotational axis of the geographical north pole. The variance between "true" north and magnetic north is called the *magnetic declination* or simply the *declination*. Charts of declination can be used to correct a compass to read true north in a particular location.

The magnetic field around the earth, as with a bar magnet, tends to run parallel to the surface near its equator and curves toward the earth as it moves toward the poles. It becomes vertical at the poles. Thus, at the *magnetic equator*, the lines of flux are horizontal and the angle of inclination is zero. At the magnetic poles the field is vertical and the magnetic inclination is 90° from the horizontal. A *dip needle*, which consists of a magnetic needle free to rotate on a horizontal axis, is used to chart the magnetic inclination at the earth's surface.

The strength and direction of the earth's magnetic field also varies with time. Very small variations occur daily and there are small annual fluctuations. Records show that large variations occur over centuries of time. During so-called *magnetic storms* very large, erratic variations occur. It appears that magnetic storms correlate with the sunspot cycle and are thought to be caused by electrical current anomalies in the atmosphere.

The earth's magnetic field is believed to originate from electrical currents within the earth, possibly its molten core. The theory that the earth consists of solid material forming a permanent magnct has been dispelled because of the extreme temperatures within the earth's core.

Most people are familiar with the use of the earth's magnetic field in navigation, but it also plays a role in protecting the environment. High-energy, charged particles impinging on the earth from outer space and solar eruptions are potentially dangerous to life. Because the magnetic field of the earth, for the most part, is parallel or near parallel to its surface, incoming radiation strikes it at or near right angles and is reflected away from the earth. As you move closer to the magnetic poles, the inclination becomes less, until, at the magnetic poles, it becomes zero and the incoming radiation is parallel to the magnetic field. Thus, incoming radiation can enter the earth's atmosphere at the magnetic poles, causing it to ionize and give off light. The result is the spectacular *aurora borealis* (northern lights) in the northern hemisphere and the *aurora australis* (southern lights) in the southern hemisphere.

Magnetic Permeability

The ability of a magnetic field to pass through a given material is that material's *permeability.* Permeability appears to be an affinity for magnetism. For example, a magnetic field is more dense in passing through a piece of iron than through an equal volume of air. As a result, iron is said to be more permeable than air. Permeability is expressed mathematically as the ratio of magnetic flux density to the magnetic field strength or:

$$\mu = \frac{B}{H}$$

where μ is the permeability in webers per amp-turn-meter, B is the magnetic flux per unit area and H is the magnetic field strength in amps per meter.

The relative permeability is the ratio of the permeability of a substance to the permeability of a vacuum. The permeability of air and non-magnetic substances is very close to that of a vacuum and is very close to one, or *unity*.

Permeability is very important in the construction of efficient electrical machines. Large air gaps, for example, decrease the strength of a magnetic field, thereby causing a reduction in rotational energy derived from the machine. Materials used as the core for electrical magnets and transformers are chosen for their ability to provide maximum magnetic flux. Supercooling magnetic materials substantially increases their permeability, resulting in electrical machinery that is more compactly designed and requires less energy in relation to its output.

Electromagnetic Induction

Oersted established the concept that a moving charge had an associated magnetic field. However, no one had at that time demonstrated the generation of an electric potential by a magnetic field. Eleven years after Oersted's discovery, Michael Faraday in England and Joseph Henry in the United States discovered than an electrical potential is generated in a conductor located in a changing magnetic flux. To gain a first-hand understanding of the Faraday/Henry discovery, do the activities in Experiment 13–4.

*****

Experiment 13–4. Electromagnetic Induction

Wrap about 25 turns of insulated wire loosely around the first three fingers of your hand such that you can readily slip your fingers out. Fasten the bare ends of the wire to a galvanometer (Fig. 13.6). Quickly thrust a bar magnet back and forth through the wire coil and observe the needle of the galvanometer. Did the direction of travel of the bar magnet affect the direction of deflection of the needle on the galvanometer?

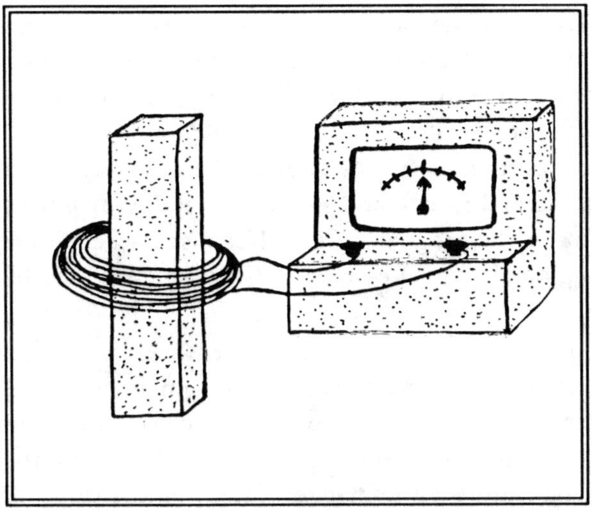

Fig. 13.6. Moving a bar magnet in respect to a copper coil.

Hold the magnet vertically on a table top and move the coil back and forth over the magnet. How does moving the coil affect the galvanometer compared to moving the magnet in the coil? How does the speed of moving the coil or the magnet affect the intensity of the galvanometer?

Move the magnet in the coil from side to side without moving it at right angles to the coil. Did this action affect the galvanometer?

Hold two bar magnets side by side with their north poles together and move them back and forth through the coil? Did this change the deflection on the galvanometer compared to using one magnet? Decrease the number of turnings in the coil to half the original number and move one bar magnet back and forth through the coil. How does this affect the deflection of the galvanometer?

You have demonstrated the generation of an electric current by moving a magnetic field through a conductor. Because no direct conduction was involved, the current was *induced* into the coil. Whenever a magnetic field moves through a conductor, an electric potential is generated.

*****

In Experiment 13–4 you observed that a magnet plunged through a coil of wire generated an electric potential, often referred to as an *electromotive force* (emf). When the magnet was pulled out of the coil, an emf of opposite polarity was developed. Moving the coil back and forth over a fixed magnet gave similar results. So, one observes an emf generated in a conductor irrespective of which is moving, as long as one moves in respect to the other to produce a changing magnetic flux in the conductor. Movement of the conductor parallel to the magnetic field produces no emf. Maximum emf is attained when the conductor moves perpendicular to the flux in the magnetic field. Because the emf was generated without making an electrical connection, it was generated by *magnetic induction.*

The cause of the emf is the reverse of the effect observed in the Oersted experiment, in which moving charges set up a magnetic field that repelled the magnetic needle. In the case of magnetic induction, the conductor contains free electrons. As the free electrons in the conductor move through the magnetic field (or the magnetic field moves through them), a repulsion is set up, and because the free electrons are not tightly bound by the atoms in the conductor, they are repelled to one end of the conductor. One can determine the polarity of the emf by using a "right-hand rule." The thumb and forefinger are extended at right angels to each other in the same plane and the index finger is placed at right angles to the plane of the thumb and forefinger. By placing the thumb in the direction of motion of the conductor and the forefinger in the direction of the magnetic field (N to S), the index finger will point in the direction of the current flow.

If a complete circuit is connected to the conductor moving through a magnetic flux, a current will flow in the conductor. This current will also generate a magnetic field, which will oppose the magnetic field through which the conductor is passing, thereby causing a counterforce. You have probably experienced this counter-force in turning the crank of an electrical generator such as in the old-fashioned wall telephones. The German physicist Heinrich Lenz first observed the counter-effect of an induced current. It is expressed in what is now known as *Lenz's Law:*

> *The magnetic effect of an induced current is in a direction which will oppose the change producing the current.*

Thus, if the direction of a conductor moving across the magnetic flux is down, the opposing force will be up, due to the induced current.

The emf induced in each turn of a coil is inversely proportional to the time rate of change of the magnetic flux and directly proportional to the change in total flux in each turn. The total emf (E) induced in a coil of N turns can be expressed by the relationship:

$$E = -N \frac{\Phi}{t}$$

where E is in volts, Φ is in webers, t is in seconds and the minus sign indicates the direction of the emf, which is opposite to the change in the magnetic flux that produced it.

Electric Generators

Electromagnetic induction is the basis for most of the electrical energy used in the world today. Electrical generators, known as *alternators,* are used in automobiles, hydroelectric plants, thermo-electric plants and nuclear power plants. All use the principle of electromagnetic induction. An alternator contains an *armature*, which consists of coils around a rotating wheel that passes the poles of strong electromagnets, generating an emf in the coils. The emf produces a current that is conducted through metal brushes riding on *slip rings*. These are mounted via insulators on the rotating shaft that connects the armature to the outside terminals (Fig. 13.7). Because the coils at one time pass through the magnetic field in one direction and then, as the shaft rotates, pass through it in the opposite direction, the net current produced is an *alternating current* (AC).

An alternating current builds from zero to a positive peak, decreases to zero, reverses direction, builds to a negative peak and then decreases to zero. In household current, for example, the process repeats itself sixty times each second of time. Figure 11.5 (Chapter 11) is a representation of such a wave.

An alternator with a *split-half commutator* attached to its shaft (Fig. 13.8) produces a *direct current* (DC). The split halves, which are insulated from each other and the shaft, are positioned so that, as the armature reverses direction in the magnetic field, they make contact with the opposite brushes. Thus, the current is always sent to the terminals in the same direction, producing DC.

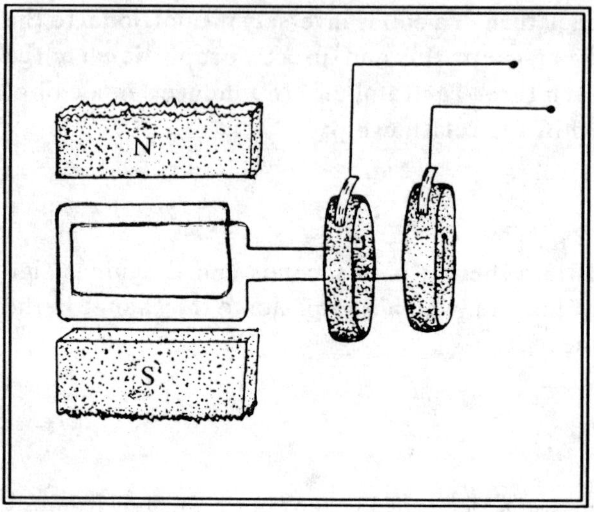

Fig. 13.7. An AC alternator.

Direct current generators have lost their popularity because of problems associated with the commutators as well as long-distance transmission losses associated with DC. Even the automobile generator has been replaced by the alternator because of the development of economical solid-state devices for converting AC to DC for charging the battery. An advantage of the alternator is that slip rings are easier to maintain than a commutator.

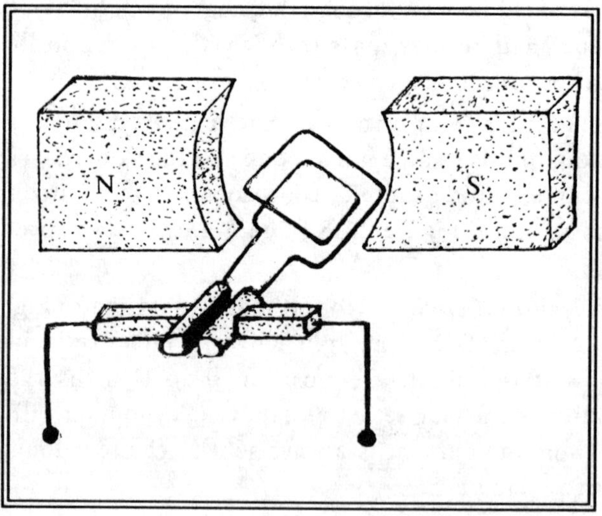

Fig. 13.8. The DC generator.

Transformers

Due to the resistance inherent in power lines, the long-distance transmission of DC electricity is not practical. The device that made alternating current appealing was the *electrical transformer*. A transformer consists of an iron core and two or more windings superimposed over one another. Figure 13.9 is drawn with the windings separated for clarity. The input winding is called the *primary* and the output winding is the *secondary*. An alternating current in the primary sets up an alternating magnetic field in the core, which results in a changing flux in the secondary. An alternating current is thereby induced in the secondary. If the number of windings in the secondary equals the number of windings in the primary, the potential in volts will be the same across each. However, if the secondary has more windings than the primary, the voltage in the secondary will be greater, resulting in a *step-up* transformer. Conversely, more windings in the primary than in the secondary results in a *step-down* transformer, with a resultant lower voltage in the secondary.

Transformers have many applications. After many miles of transmission, the voltage of an electrical current drops due to resistance loss in

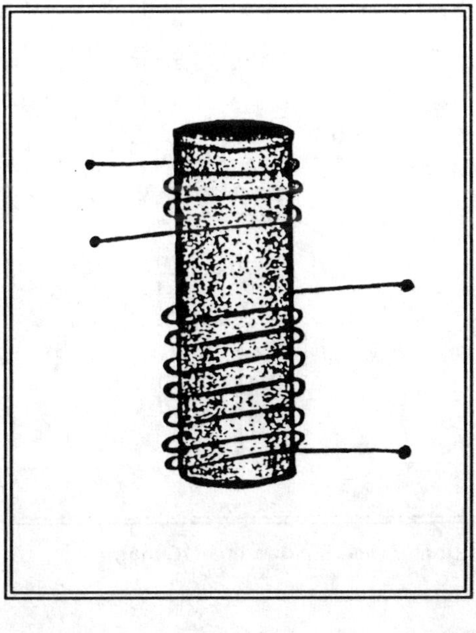

Fig. 13.9. A step-up transformer.

the lines. A step-up transformer raises the voltage to the desirable level. The voltage in most power lines is much too high for home use, and a step-down transformer is placed near your home to produce the 115–220 volts necessary for home appliances. A high voltage is needed at the spark plug of a gasoline engine to ignite the fuel-air mixture (Chapter 6). The device which produces the high voltage is a step-up transformer called an *induction coil.* The DC from the battery is converted to the AC necessary for the operation of the induction coil.

Electric Motors

Electric motors can be designed for DC, AC or AC/DC. All produce rotary motion through the principle that unlike magnetic poles attract and like magnetic poles repel.

The DC motor consists of an armature (electromagnet) mounted on a shaft, which is free to rotate in the field of a permanent (field) magnet.

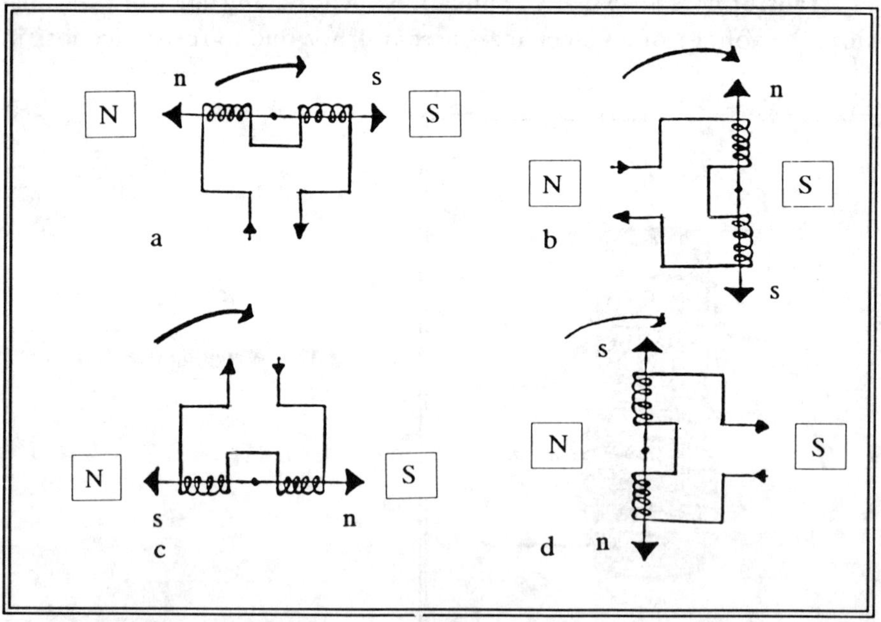

Fig. 13.10. A schematic of the rotation of the armature in a DC motor.

Each of the leads of the armature is connected to a respective segment of the commutator (see Fig. 13.8 for commutator detail) that reverses the current polarity of the armature as it rotates in the field. Initially, the current flowing through the armature magnetizes it with the same polarity as the field (Fig. 13.10a). The armature is repelled at both ends and rotates one-fourth revolution where the north pole begins to be attracted by the south pole of the field (Fig. 13.10b) and the south pole of the armature is attracted by the north pole of the field. In Figure 13.10c the current is reversed in the armature, which gives it the same polarity as the field. It is repelled a quarter-turn and then attracted by the opposite polarity of the field (Fig. 13.10d). The process continuously repeats itself, causing smooth rotary motion.

A small DC motor can also be operated on AC because the current polarity in both the armature and the field reverses simultaneously. Consequently, the rotation continues in the same direction. Special design is necessary, however, due to considerable heat generation in the windings.

The induction motor, which is the most common AC motor currently in use, consists of field coils arranged around a rotor made of laminated iron sheets with slots containing interconnected copper bars. The alternating current sets up an alternating magnetic field that induces an electric current in the rotor. The resultant alternating magnetic field in the rotor is repelled to produce rotary motion.

The Propagation of Radio Waves

The first radio-wave transmission developed by Marconi consisted of a series of electrical spark discharges arranged in Morris code. An electrical discharge gives off an electrical pulse, such as you've heard on a radio when you operate an electrical switch. Radio waves are electromagnetic, thereby consisting of an electric component and a magnetic component. James Clark Maxwell, a Scottish physicist, developed a mathematical model for electromagnetic waves that consist of a magnetic field at right angles to an electrical component traveling in a direction perpendicular to both fields. Electromagnetic waves sustain themselves according to Faraday's law. A changing electric field generates a perpendicular magnetic field, which in turn generates a perpendicular electric field. The cycle is continuously repeated and therefore, no medium is required for propagation. All electromagnetic radiation, such as radio waves and X-rays, travel at the speed of light of 3×10^8 m/sec. The electromagnetic spectrum consists of radio waves (including television), microwaves, infrared light, visible light, ultraviolet light, X-rays and gamma rays, in order of their ascending frequencies and descending wavelengths. The physical nature of the radiation is the same throughout the spectrum. Electromagnetic radiation varies only in frequency and wavelength.

Radio waves are generated by feeding an alternating potential to an antenna at a frequency and energy level sufficient to cause some of the energy to break away. The resultant electromagnetic radiation travels through space and generates a varying potential of the same frequency in the receiving antenna. This varying potential is fed to a radio receiver, which converts it to sound waves. Each radio transmitter sends out a specific, assigned frequency that can be "tuned" in by a radio receiver.

Originally, radio signals were "detected" by diodes of semiconducting crystals of *galena* (lead sulfide) and a brass "cat whisker" (brass wire) touching its surface. The feeble current from the received radio signal would flow through them only in one direction (rectifying it), resulting in a uni-directional signal which would activate headphones. The vacuum tube diode, or "Fleming valve," invented by J.A. Fleming, replaced the crystal diodes and led to the development of the vacuum radio tube.

Vacuum Radio Tubes

Thomas Edison observed dark spots on the bulb of his burned-out light bulbs. He placed a metal plate with an external lead above the filament inside a light bulb. When a galvanometer was connected between the filament lead and the plate lead, a current was observed. Edison made no application of his observations, as he was interested only in solving the problems of burned-out filaments.

The *Fleming diode* (two elements) made application of the effect observed by Edison. A hot filament emitted electrons, which were attracted by a positively charged plate. When the input signal was positive, the plate (*anode*) attracted electrons from the negatively-charged *cathode*

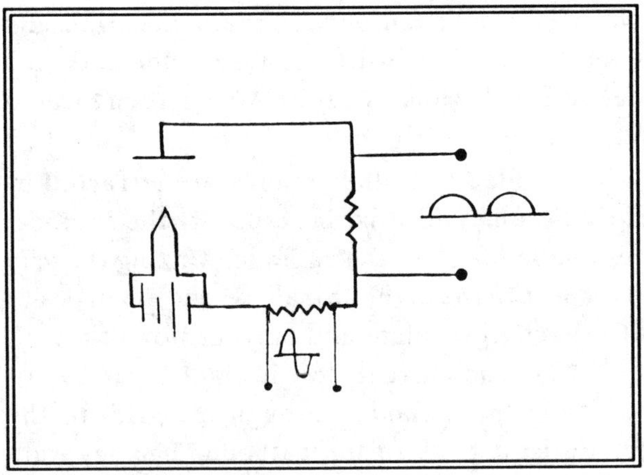

Fig. 13.11. A diode amplifier circuit.

and current flowed through the output resistor (Fig. 13.11). The negative portion of the wave turned the plate negative in respect to the filament and current did not flow. Only the positive portion of the wave appeared in the "rectified" output. Consequently, diodes are often called rectifier tubes. They still have application in rectifying AC into DC, although solid-state rectifiers have replaced them to a considerable degree.

The principle of electrons being emitted from a hot filament is known as *thermionic emission*. Heat energy imparts sufficient energy to the free electrons in the filament to allow them to escape from its surface. Because the filament, which emits electrons, is the negative terminal, it is referred to as the *cathode*. Modern cathodes are separated from the filament and "indirectly" heated by it (Fig. 13.12).

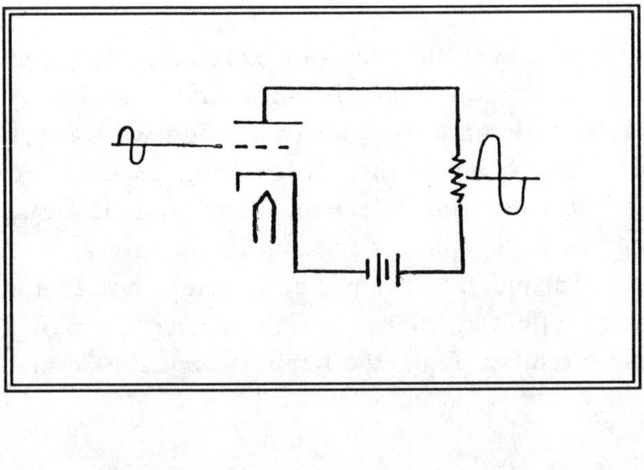

Fig. 13.12. A triode amplifier circuit.

The vacuum tube, which launched radio-wave communication and sound amplification, is the triode. It consists of three elements, the cathode, the plate (anode) and grid, which is a spiral wire screen located between the cathode and plate.

As in the diode, electrons boiled from the cathode are attracted by the positive plate. If the grid becomes negative in respect to the cathode, the emitted electrons are repelled back to the cathode. Making the grid less negative (positive in respect to the cathode), allows the electrons to be attracted by the positively-charged plate and current flows through the output resister (Fig. 13.12). The current flow in the output circuit moves through the triode from the cathode, through the grid, to the plate, through the output resistor back to the cathode. The separate filament, which heats the cathode, is connected to a separate circuit.

The grid in a triode, is essentially a gate in the flow of electrons from the cathode to the plate. When operating as an amplifier, the grid operating voltage, or *bias voltage*, is pre-established, so that the cathode to plate current never reaches zero. Therefore, the plate current fluctuates around the grid voltage. When the wave, applied to the grid turns positive, the plate current is high and when negative, the plate current is low. The plate circuit current is very high, which results in a very high potential across the output resistor in the plate circuit. Consequently, a small wave on the grid results in a very large wave on the output resistor (Fig. 13.12), resulting in high amplification. Triodes in tandem lead to very high amplification capable of driving very powerful loudspeakers.

The Cathode Ray Tube

Cathode ray tubes form images in television sets and computer monitors. The basis for the cathode ray tube is thermionic emission. A heated cathode emits electrons that are attracted by an anode with a very high positive potential at the front of the screen (Fig. 13.13). The electrons are converged into a very fine "electron beam" by a magnetic focusing coil and impinge on a phosphorus screen, causing a very fine spot on the screen to glow. The electron beam is deflected by horizontal and vertical electrostatic deflecting plates, which receive a varying potential from electronic circuitry. Thus, the beam sweeps the screen

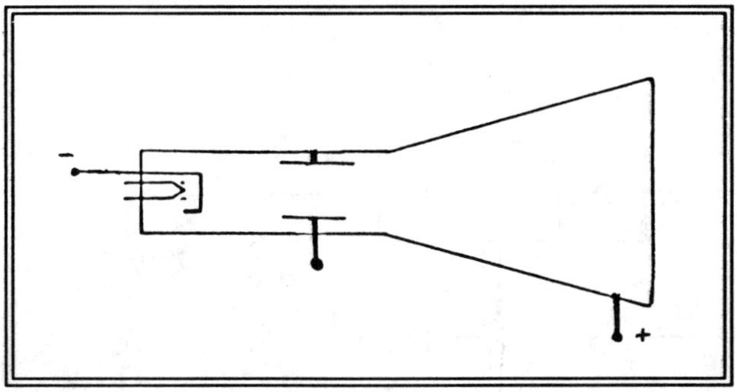

Fig. 13.13. A cathode ray tube.

vertically and horizontally, resulting in a completely filled screen. If you look at a large television screen with the tuner off-station so no image appears, you can see the individual "traces" of the electron beam. As the beam sweeps the screen, the intensity is varied by the amplified incoming signal, causing an image on the screen.

Transistors

Certain materials act as *semiconductors*. The earliest semiconductors were diodes, consisting of a fine platinum wire anode fastened to a crystal of the element germanium. This diode acted as the Fleming valve in producing rectified current but without heat loss and electrical circuitry. Thus, semiconductors are more compact and efficient than vacuum tubes and eliminate considerable heat, which can damage electronic equipment.

Transistors can also function as amplifiers by forming two electron donor-electron acceptor junctions, referred to as *junction transistors*. To produce amplification, it is necessary to control the output current with the input signal, just as with the triode vacuum tube. The two junctions in the transistor are, in effect, equal to opposing diodes in series.

The transistor consists of two layers of like material separated by a different material. The materials are either electron donors or electron acceptors. The element germanium, by itself, is practically an insulator, but when carefully "doped" with controlled amounts of arsenic, it forms an electron donor material referred to as *N-type*. When doped with boron, it forms an electron accepting or *P-type* material. Bonding P-type

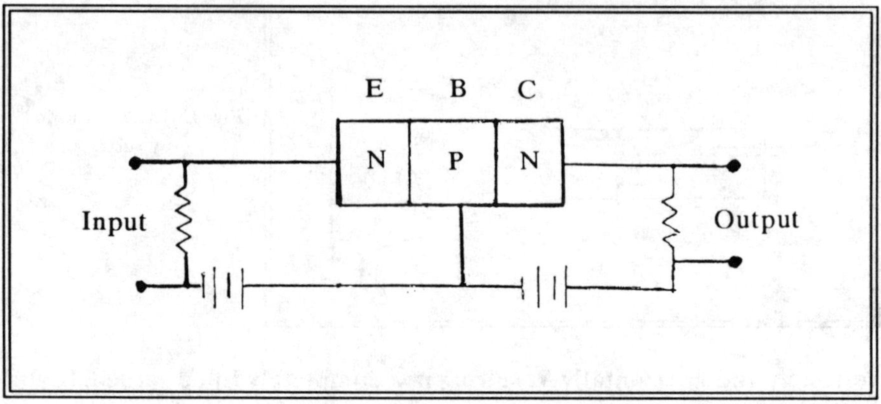

Fig. 13.14. N-P-N amplifier circuit.

material between two pieces of N-type material results in a *N-P-N-transistor*. Sandwiching an N-type material between two P-type materials yields a *P-N-P transistor.*

Figure 13.14 illustrates an N-P-N transistor with emitter (e) on the left and collector (c) on the right, separated by the base (b). The base serves as a barrier to current flow from the emitter to the collector, which is controlled by the bias current set up by the battery in the input circuit. Keep in mind that electrons can always flow from N to P material across an NP junction, but cannot flow from P to N under normal conditions. Therefore, in Figure 13.14, the current can flow from the emitter to the base and is said to be "forward biased" in the emitter circuit. However, current cannot flow in the collector circuit, because it is "reverse biased."

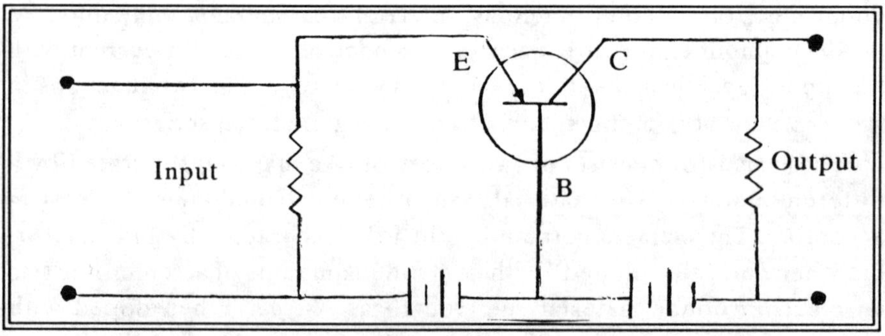

Fig. 13.15. P-N-P amplifier circuit.

A signal in the input circuit at one point in time will override the emitter bias. This causes more free electrons in the base and, in effect, reduces the barrier effect at the P-N junction, letting current flow to increase from emitter to collector. The current flow from the emitter through the base and the collector develops a potential across the output resistor. The opposite effect is observed when the input signal swings in the opposite direction. As with the triode vacuum tube, a small input signal is able to control a large output current to provide amplification. Figure 13.15 illustrates a P-N-P transistor amplifier circuit in schematic form.

Transistors have wide and varied uses. Due to their compactness, high efficiency, dependability and relatively low production cost, they have practically replaced vacuum tubes. Vacuum tubes are relatively large and require filaments that often have problems with heat dissipation, which causes a short life. Some of the original vacuum-tube mainframe computers required huge cabinets for housing and elaborate ventilation systems to carry away the heat. The majority of down-time was due to vacuum tube failure. Such things as space travel communication and personal computers would not be possible were it not for ultra-small transistors integrated with circuitry into "microchips."

Science/Technology/Society

Electromagnetic induction is employed in a variety of ways in our modern society. Electric motors are used in just about every conceivable manner. Some are quite necessary in an industrial society, but others are really just luxuries.

Electromagnetic induction also makes possible the storage of audio and video information, both on plastic disks and magnetic tape, such as those used in a computer. Electrical energy can be generated by coupling heat engines and water turbines to electric generators.

While water may be considered a clean, renewable resource for energy generation, dams on rivers alter the ecology of those bodies. Although they may reduce property loss due to flooding, dams can create abrupt changes in water levels. Such changes affect the spawning and egg hatches of fish. Certain species, such as the Missouri River Paddle Fish, were seldom seen prior to the installation of dams. Now, they are routinely snagged by fishermen during the season. The fate of some species, such as the Missouri River Sturgeon, is uncertain because of the lack of knowledge about their spawning characteristics.

Power lines from generating stations and associated hardware utilize fertile land areas and affect the aesthetics of the land. Concern has developed about the electromagnetic radiation given off from high voltage power lines. Just as a magnetic field was formed around the wire when you ran a current through it over the compass, so too, a pulsating magnetic field is formed around power lines. Due to their high voltage, these fields can become quite large and powerful and some of the radiation is transmitted into the atmosphere. You can perceive this radiation as it interferes with the reception in your automobile radio as you pass under the power lines.

Magnetic radiation is used in the analysis of the treatment of certain physical malfunctions. On the other hand, there is much that needs to be learned about the physiological effects of electromagnetic radiation.

TEACHER'S GUIDE

Overview

This program is designed for people preparing for the teaching profession or who are already teaching elementary school. The major objective is to help students develop an appropriate model of science while they compile a physical science knowledge-base that will enable them to use scientific processes in teaching science and other subjects.

The overarching approach to teaching from this book should be one of inquiry and problem-solving. Science is a dynamic process that involves observing, formulating hypotheses, collecting data, using time-space relationships, interpreting data, drawing inferences and communicating findings, especially in written form. The students should be encouraged to practice these processes. Where at all possible, new information should be presented through experimentation rather than lecturing. This does not imply that students will be learning through sheer discovery. Rather, with the assistance of the teacher, they will perform carefully structured experiments that will allow them to reach plausible solutions to problems. Whenever possible, enough equipment and supplies should be made available so that students may work in pairs. Larger groups usually end up with one or two students manipulating the experiment and the others merely observing.

In most instances replication of the experiments by each pair of students to account for experimental error is not possible because of time constraints. However, data obtained by the entire class may be substituted for replication data. Class data should be presented to the whole class. Although there are problems associated with this method (for example, there are bound to be errors and omissions because it is the students' first experience with the activity), when class data are displayed on the chalkboard or via an overhead projector, patterns and anomolies do emerge from which the students can draw appropriate conclusions.

Your foremost concern should be for the safety of your students and the use of appropriate procedures that decrease the probability of injury to zero. As the teacher, you must always anticipate hazardous situations. Approved laboratory safety procedures must be followed at all times during the experiments described in this textbook, and a qualified person must be in the laboratory when experiments are performed. Students must never be left on their own doing experiments.

A model of the apparatus should always be set up by the instructor prior to the experiment. The instructor should check and approve each student's apparatus prior to allowing them to begin the experiment. One needs to be especially observant in this process, as such unlikely things as substituting a solid glass rod for glass tubing, thereby causing an explosion, do occur. Safety precautions should be reiterated for each experiment. It is not enough to verbalize these precautions at the beginning of the term of study.

THE ACTIVITIES IN THIS TEXT MUST BE SUPERVISED BY AN EXPERIENCED, QUALIFIED TEACHER OF PHYSICAL SCIENCE.

Equipment, Chemicals and Supplies

The following is a list of materials that will be used by the students during the entire course. Specific supplies for each experiment are listed under that experiment.

<u>Quantities per student station</u>

1 pegboard apparatus
2 small test tube clamps
2 large test tube clamps
5 small test tubes (20 mm x 150 mm)
5 large test tubes (25 mm x 150 mm)
1 double pan balance (non-damping)
1 set of masses
1 250 mg mass
1 plastic box with 250 beads (0.2 g) from bead chain
1 electrolysis kit (electrodes)
1 burner stand (for alcohol burners)
1 alcohol burner (two-position type)
1 elasticity of gases apparatus (syringe in wooden block)
1 density of solids kit: 2 identical-sized cubes of distinctly different masses, a rectangular slab and a 30-cm plastic ruler

1 thermal expansion of solids kit: metal tubes, glass tube, corsage pin, clothes pin with mounting hole, mounting screws, cardboard 360° dials, masonite blocks with emery paper, corks and thumb tacks
1 elasticity of solids apparatus
1 250-ml pyrex beaker
1 400-ml pyrex beaker
1 alcohol burner (2 position)
1 metal pie pan & sand for alcohol burners
1 glass funnel
2 soda straws
1 evaporating dish
2 laboratory thermometers
1 glass stirring rod
1 evaporating dish
1 set wave-demonstration springs: one 3" dia. x 4" to 6" long, and one 6 ft. long x approximately 1" dia.
1 optics kit: plane mirror, concave mirror, plastic sandwich box (semicircular refracting), glass lens (convex), wood block, protractor and rubber bands
3 miniature socket and screw-type flashlight lamps
1 concave lens
1 optics slits kit: 2 microscope slides, graphite suspension, 2 double-edge razor blades, showcase lamp, socket, cord & plug and rubber bands.
1 diffraction grating mounted in a 2" x 2" slide holder
1 vinyl plastic strip, 3 cm x 30 cm
1 acetate plastic strip, 3 cm x 30 cm
monofilament nylon fishing line to support plastic strips
2 styrofoam spheres, about 2 cm dia.
1 spectral analysis kit: 1 vial each of calcium chloride, copper chloride, copper nitrate, lithium chloride, potassium chloride, potassium nitrate, sodium chloride, bicarbonate of soda, strontium chloride, strontium nitrate and 10 nichrome wires with balsa wood handles.
1 machine bolt, 3/8" x 2"
5 tacks, small
2 baby food jars with lids
1 magnetic compass
1 20-cm piece of rubber tubing
2 50-ml graduate cylinders
2 wide rubber bands and 2 narrow rubber bands
2 Alka-Seltzer tablets
1 plastic teaspoon
1 5-qt plastic ice cream bucket
1 centripetal force apparatus: braided fishline, masking tape, large paper clip, 5 large washers, #4 two-hole rubber stopper, 15 cm x 10 mm diameter glass tubing, firepolished on both ends and wrapped with a layer of plastic tape.
1 soda straw vane electroscope
2 size *D* cells and holders
2 bar magnets

One source for the special equipment illustrated in the photographs and referenced in the textbook is the curriculum supply section in Frey Scientific Company catalog, Frey Scientific, 905 Hickory Lane, P.O. Box 8101, Mansfield, OH 44901-8101.

General Materials Needed

Alcohol burner fuel
Alum
Boiling chips
2 dropping bottles containing glycerine
Single-cylinder lawn mower engine (scrap)
Filter paper
Sand, fine, dry and clean
Black ink, one bottle
Food coloring - two bottles, each of a different color
Naphthalene
Dichlorodifluorobenzene
Methanol
Isopropyl alcohol
Glycerine
Magnesium turnings
Limewater
Sulfuric acid
Hydrochloric acid
Nitric acid
Ammonia water
D.C. power supply, 6-12 volts
Interesting juices materials: bromothymal blue, distilled water, orange juice, lemon juice, a cola, limewater, milk of magnesia, vinegar, ammonia water, tap water, lime juice, Draino, 2 hair shampoos, Alka-Seltzer, dilute sulfuric acid, dilute nitric acid, dilute hydrochloric acid
Sodium carbonate
Wood splints
Redwood splints for destructive instruction
Sodium chloride
Potassium nitrate
Sodium nitrate
Sodium chlorate
Charcoal
Nitric acid
Copper oxide
Galvanometer
Zinc metal
Mercury vapor lamp

General Safety Equipment

The following safety equipment should be available for the students. Its use should be required any time an experiment might involve the student in a hazardous situation.

Eye wash
Fire blanket
Goggles: must exhibit the ANSI Z.87 logo
Pan and sand for alcohol burners
Safety hood

****

Chapter 1

Experiment 1–1. Educational Destruction

Objective: To develop the concept of breaking materials down into constituent parts for analysis.

Problem Statement: Is wood more than one substance?

Supplies:*
alcohol burner
pegboard support
2 test tube clamps
1 plastic bucket
wood splints
1 large test tube with 2-hole stopper
right-angle glass tubing
glass tubing
1 small test tube
2 baby food jars with lids
matches

Safety Precautions: This is the first experiment the students will do and it provides a mechanism to set the stage for becoming familiar with science processes and the assembly and operation of rudimentary science equipment. Because the students are unfamiliar with the equipment, the physical layout of the room and other class members, there is a greater potential for accidents and injury. Therefore, following safety precautions is particularly important.

---

* All supplies listed for experiments are for each student station.

Goggles must be worn. Because the right-angled glass tubing is brittle and can break while being pushed into the stoppers, the stopper holes should always be lubricated with glycerine. The tubing should be grasped close to the end, rather than at the elbow, and a terry cloth towel segment should be placed between it and the hand for protection. Hot glass appears the same as cold glass and care should be taken in handling the large test tube after heating. A test tube should always be clamped next to the flange at the mouth and the clamp should not be heated. If the spring-type test tube clamps, as shown in the figures, are used, only large clamps should be used on large test tubes and small clamps on small test tubes to reduce instability. Alcohol lamps pose special problems as alcohol flames are colorless and hard to see. The burners should always be used in pie tins of sand, so if upset, the alcohol is contained. Students must never be allowed to fill alcohol burners and the alcohol must be stored in an approved flameproof container. One should not try to ignite the gas coming from the delivery tube, as there is the possibility of the flame "backing into" the generator, resulting in an explosion. The delivery tube should always be taken from the water prior to removing the flame from the generating tube as reversing the process causes the tube to cool and the atmospheric pressure to force the water into the hot generating tube and breaking it. Stressing these safety precautions from the first will set a precedence for the use of these procedures in future experiments.

Procedures: Follow the procedures given in the text. The trap should be placed in very cold water and the gas collected in baby food jars, or similar containers. The materials can be examined visually and the gas ignited to show a blue flame. The "charcoal" can be tapped out of the large test tube and burned to give off "smoke" and an ash remains. It is not necessary to identify the products, but the students should understand that the wood was heated and several products were given off.

Redwood splinters seem to work best for the generating tube. They are easily split from scraps of redwood siding, shakes or one-inch pieces. The students should break the splinters short enough so they clear the one-hole stopper and do not inhibit it from seating firmly into the generating test tube.

****

Chapter 2

Experiment 2–1. Air in Sand

Objective: To determine the amount of air contained in a sample of sand.

Problem Statement: What percentage of sand is air?

Supplies: 2 50-ml graduated cylinders
Sand

Safety Precautions: Follow normal laboratory safety procedures.

Procedures: The sand should be fine-grained and dry. The students should be certain that the water seeps completely through the sand. Therefore, it is best to pour the sand into the water. The graduate cylinder containing the sand must be dry, so that sand particles do not stick to the graduate while being poured into the water. Experimentally, you should expect that about 35% of the sand is air.

*****

Experiment 2–2. The Double Pan Balance

Objectives:
1. To determine mass using a double pan balance.
2. To develop graphing skills.
3. To develop the concept of experimental error.

Problem Statement: How is the mass of an object determined using a double pan balance?

Supplies: 1 double pan balance
1 200-mg mass
1 set of beads

Procedures: It is best if the balances do not have calibrations on them so the students may set up their own calibrations with pencil marks. After explaining the operation of the balance, the students should "zero-in" the balances prior to determination of mass. The 200-mg masses should be distributed. The teacher may need to convert the milligrams to grams for the students, as many will not know how to do it. Placing the mass on the double pan balance will allow the students to calibrate their balances. If more than one section of the class is taught, or if calibrated balances are used, the students should verify the markings.

Bead-chain that has a mass of about 0.2 g/bead (or multiples of 0.1 g) should be used. The mass of one bead should be determined graphically, using dots with circles around them to indicate experimental error.

*****

Problems

1. 2 cm
3. The same
5. 14.9 g
7. 2500 cm^3
9. 0.18 g/cm^3
11. Cut one cubic centimeter of the ice while in an environment where the temperature is maintained at the freezing temperature of water and determine its mass. Water expands when it freezes and so the mass will be less than that of liquid water.
13. Capture all of the products of burning and determine their mass.
15. 1.0×10^{-3}; 5.0×10^{-4}; 2.5×10^{-2}; 5.2×10^{-4}
17. 5.428×10^{3}; 9.3×10^{7}; 6.21×10^{2}; 4.782×10^{3}
19. 2.5×10^{-3}; 4.1×10^{-1}; 8.567×10^{-1}; 9.2×10^{-7}

****

Chapter 3

Experiment 3–1. Density of a Solid

Objective: To determine the mass densities of a regular and irregular solid.

Problem Statement: Is the mass density of a solid a characteristic property?

Supplies:
2 metallic cubes, each about two centimeters on edge, but of distinctly different densities.
1 rectangular, non-cubic solid of the same density as one of the other cubes
1 set of masses
1 set of beads
1 double pan balance
1 graduated cylinder (50 ml)

Safety Precautions: Follow normal laboratory safety precautions.

Procedures: Measurements with the plastic rules should be done with the edge of the rule along the edge of the solid. If the rules have no specific zero mark, students should use the 1-cm mark as the reference point and add one centimeter to their measurements. Class data will fall within close agreement;

however, the densities of the pebbles will vary because different kinds of pebbles will be used. The instructor should dispel the common myth that the size of the pebbles causes the variations in densities!

*****

Experiment 3–2. Density of a Liquid

Objectives: To determine the density of a liquid.

ProblemStatement: Is density a characteristic property of a liquid?

Supplies:
1 double pan balance
1 set of masses
1 set of beads
500 g of Epsom salt
2000 ml of water

Safety Precautions: Goggles required.

Procedures: Dissolve 400 grams of Epsom salt in 1000 ml of water for one solution and 100 grams in 1000 ml of water for the second. Students need to be instructed in reading a graduated cylinder correctly.

*****

Experiment 3–3. Density of a gas

Objective: To determine the density of a gas.

Problem Statement: Is density a characteristic property of a gas?

Supplies: Alka-Seltzer tablets

Safety Precautions: Goggles required for this experiment. Follow normal laboratory safety procedures.

Procedures: Alka-Seltzer tablets are used to generate the gas. The tablets need to be broken in half to fit into the large test tube generator prior to mass determinations. Only the particles placed on the balance will be used in the generator.

The glass tube should just peek through the stopper in the generating tube so no liquid goes into it. The reaction begins rapidly, so the students need to move quickly and deliberately. Collect gas for at least ten minutes to make sure most of the gas, carbon dioxide, given off is collected.

At least a 500-ml jar is required to collect the gas. No air bubbles should be in the jar when it is filled with water and the delivery tube should be placed near the top of the jar as enough of the carbon dioxide gas can dissolve to give erroneous readings.

*****

Problems:

1. (a) The same as the graph in Figure 3.8a; (b) horizontal line.
3. 0.45 g
5. 3,240 g
7. (a) Density; (b) Thermal expansion, density; (c) All three
9. (a) A = 19.3 g/cm^3; B = 11.3 g/cm^3; B is greater. (b)8.0 g/cm^3
11. 9.65 cm^3
13. 3,396.8 pounds
15. 10.2 cm

****

Chapter 4

Experiment 4–1. Centripetal Force

Objective: To determine that centripetal force is directly proportional to the mass and velocity, but inversely proportional to the radius of curvature.

Problem Statement: How are the velocity, radius and radius of curvature related to the centripetal force accelerating a rotating body?

Supplies:
1 # four, two-hole stopper
2 meters fishline (do not use monofilament line).
1 15-cm glass tube (inside diameter 4 mm minimum)
1 large paper clip
5 large washers
1 two-hole rubber stopper
masking tape

Safety Precautions: Goggles required. The stopper should be firmly tied to the fishline so it does not come off during the experiment. The glass tubing should be fire-polished and wound with plastic electrical tape. Choose masses that do not require high-speedrevolutions of the stopper. The radius of rotation should not exceed one meter, and groups should be arranged to minimize the danger of being struck by the stoppers.

Procedures: Follow those given in the text.

*****

Problems

1. (a)sitting; (b) centripetal force; (c) fly wheels; (d) pushing a lawn mower
3. 500 nt
5. 60 m/sec
7. 225 kg
9. (a) Mud flying off an automobile wheel. (b) An automobile door pushing you into a curve as you go around a corner

****

Chapter 5

No experiments have been suggested for this chapter, but the horsepower of students running to the top of a set of stairs could be computed. A laboratory mass on the end of a cord can be used to demonstrate the conservation of energy of a pendulum as in Figure 5.5. CAUTION: You must hold your head against a solid support such as a wall, as a slight movement forward can bring your head into the mass on its return swing! The students should be cautioned, if they decide to try it.

*****

Problems

1. None
3. 70,560 joules
5. 700 joules
7. a)100 joules; b)none; c) 200 joules
9. 72.7 watts

****

Chapter 6

Experiment 6–1. Thermal Expansion of Solids

Objective: To determine the linear thermal expansion of solids.

Problem Statement: Is linear thermal expansion a characteristic property of solids?

Supplies:
- 1 peg board support
- 1 alcohol burner
- 1 large test tube clamp
- 1 large test tube
- 1 corsage pin
- 1 cardboard circular dial with 360° markings
- 1 masonite block with emery cloth glued to one surface (0.5 cm x 2 cm x 5 cm)
- 1 right-angle glass tubing
- 1 #4 one-hole stopper
- 1 rubber delivery tube
- a selection of hollow rods (e.g., aluminum, copper, and glass) of the same inside diameter and approximately 28 cm long
- 1 rod (aluminum) of larger inside diameter

Safety precautions: Goggles required for this experiment. The students should position themselves so that hot water from the open end of the rods will not hit them. The tubes will be hot and must be allowed to cool before they are changed. Replace the water in the generation test tube each time the rods are changed.

Procedure: It is important that all the tubes be of the same diameter and that an extra aluminum (or glass) tube of perceptibly larger diameter be used to observe that diameter is not a factor in linear thermal expansion. Make certain the students set up the apparatus each time so that the length of the tube between the pin and the clothes pin is the same and that the pin is zeroed in the same each time. It is a good idea to elevate the clothes pin end of the support board about one-quarter of an inch to help the condensed water flow through the tube. Also, do not allow the delivery tubes to droop below the level of the rods under test, as a water trap forms and steam will not be able to travel through the tube. Only when a healthy plume of water vapor emits from the open end of the rods will the maximum expansion take place.

*****

Experiment 6–2. Boiling Temperature

Objective: To graphically determine boiling temperature.

Problem Statement: Is boiling temperature a characteristic property of liquids?

Supplies:

1 small test tube clamp
1 large test tube clamp
1 large test tube
1 small test tube
1 #4 two-hole rubber stopper
1 pegboard support
1 alcohol burner
1 burner support
1 rubber delivery tube
1 beaker, 250 ml or larger
1 thermometer
boiling chips
Isopropyl alcohol, methanol, enthanol and water

Safety Precautions: Goggles required. Students should be cautioned to use a water bath and not heat the test tube directly. The cold-water test tube condenser should contain very cold water in the beaker so alcohol vapors do not enter the laboratory, as they are very flammable.

Procedures: Follow those given in the text. Three or four boiling chips should be added to allow for smooth boiling of the liquid. Isopropyl alcohol, methanol, ethanol and water samples may be given to different members of the class.

*****

Experiment 6-3. Freezing/Melting Temperature

Objective: To graphically determine the freezing/melting temperature of a liquid

Problem Statement: Is the freezing/melting temperature of a substance a characteristic property?

Supplies:

1 pegboard support
1 alcohol burner
1 burner support
1 large test tube
1 large test tube clamp
1 beaker, 250 ml or larger
1 thermometer

1 one-hole cork with groove cut in the side to allow reading the thermometer
1 small test tube clamp
paradichlorobenzene (moth flakes)
naphthalene (moth balls)

Safety procedures: Goggles required. A water bath should be used to melt the crystals in the test tube.

Procedures: Some of the students may be given paradichlorobenzene (moth flakes) or naphthalene (moth balls). Explain to the students that the experiment begins when all of the crystals have melted and the burner is removed from the water bath.

*****

Heat engines were included in this chapter because they are fundamental to the operation of the present society. A fundamental knowledge about heat engines is useful in understanding science/technology/society interactions and in teaching children about heat engines. A single-cylinder, four-stroke gasoline engine, such as that used on a lawn mower, is an excellent device for showing the operation of the gasoline engine, because the students can observe the operation of the various parts.

*****

Problems

1. 1.275×10^{-2}
3. To allow for thermal expansion of the concrete.
5. about 25°C
7. 50.042
9. 0.0012 cm
11. Yes
13. 3,000 k-cals
15. 79°C
17. Gas turbine. Gas turbine.
19. Exhaust
21. Intake
23. The heat of compression
25. Yes. Answers may vary. Usually, loud noises without rhythm.

****

Chapter 7

Experiment 7–1. Solubility

Objective: To determine the solubility of a substance.

Problem Statement: Is solubility a characteristic property?

Supplies:
1 small test tube
1 evaporating dish
1 alcohol burner
1 burner support

Safety Precautions: Goggles required. During evaporation, the burners should be moved to reduce the heat to avoid spattering.

Procedures: Half the class will use 6 g of sodium chloride (NaCl) and half will use 6 g of sodium nitrate ($NaNO_3$). Both are poured into 6 ml of water in a small test tube. Students will be able to see that the sodium chloride does not dissolvc as rcadily as either of the other two. The liquid should be decanted so no undissolved solid moves into the evaporating dish.

*****

Experiment 7–2. Temperature as a Function of Solubility

Objective: To determine the solubility of a solid in water at a specific temperature.

Problem Statement: Is solubility at a specific temperature a characteristic property?

Supplies:
sodium choride or potassium nitrate
small test tube

Safety Precautions: Goggles required.

Procedures: Half the class is given 8 g of sodium chloride and the other half is given 8 g of potassium nitrate. Each group adds their substance to 10 ml of water in a small test tube. A hot water bath should be used to heat the solution, but the water level should not be higher than the liquid in the test tubes or too much time will be spent heating the water. Once the water has come to boil, comparisons can be made as to the amount of solute that will dissolve.

This experiment is a qualitative comparison, but various students could be assigned different temperatures and use the procedures of Experiment 8–1 to plot the potassium nitrate on a solubility/temperature graph.

*****

Experiment 7–3. The Solubility of a Gas

Objective: To qualitatively determine the solubility of a common gas.

Problem Statement: Is the solubility a characteristic property of a gas?

Supplies:
1 alcohol burner
1 burner stand
1 small test tube
1 beaker, 250 ml or larger
1 one-hole stopper
1 15-cm length of glass tubing
10 ml of ammonia water

Safety precautions: Goggles required. The ammonia solution should not be breathed directly, but should be wafted with the hand and breathed cautiously. Students should be cautioned to use a piece of glass tubing in the stopper and not a solid glass stirring rod.

Procedures: Household ammonia water may be purchased locally or concentrated aqueous ammonia can be diluted to a 1:2 solution. The test tube filled with gas can be transferred to the water in the beaker by bringing it, with the stopper and tubing, next to the beaker, removing the stopper and inverting the test tube into the water. It should be left leaning against the side of the beaker for several minutes. Water should rise into the test tube fairly quickly. It should be explained that atmospheric pressure forces the water into the test tube because the pressure inside the test tube is reduced when the gas dissolves in the water.

An "ammonia fountain" makes an excellent demonstration. A model of an ammonia fountain can be found in many chemistry textbooks and demonstrations books such as Alyea, H.N. & Dutton, F.B. (1965). *Tested Demonstrations in Chemistry* (Easton, PA: Journal of Chemical Education).

*****

Experiment 7–4. Alcohols as Solvents

Objective: To determine solubility of solids in alcohol.

Problem Statement: Can generalizations be made from just one observation?

Supplies: a few crystals each of naphthalene and sugar
water and alcohol
2 small test tubes.

Safety Precautions: Goggles required.

Procedures: Very small amounts of the solutes should be used, as large chunks will saturate the solution. Water added to the alcohol-naphthalene solution will cause a coagulated, white precipitate, as that resulting from the addition of vinegar to milk. The concept of precipitate should be discussed. One would expect parallel results from adding methanol to the sugar-water solution, but it does not materialize, showing generalizations cannot be made from one example.

*****

Problems

1. 132 g
3. 12 g
5. 40°C

****

Chapter 8

Experiment 8–1. Separation of Mixtures of Solids

Objective: To separate a mixture of solids.

Problem Statement: How can two solids mixed together be separated?

Safety Precautions: Goggles required. Use caution in evaporating the filtrate. Apply heat gently to avoid spattering.

Supplies: 1 funnel
1 piece filter paper, circular
5 grams of the mixture (50% NaCl and sulfur powder)

Procedures: Follow those given in the text.

*****

Experiment 8–2. Fractional Distillation

Objective: To describe the process of fractional distillation.

Problem Statement: Can two or more liquids be separated by utilizing their boiling temperatures?

Supplies: The mixture used is made by adding 500 ml of methanol to 500 ml of water.

1 beaker
1 alcohol burner
1 burner stand
1 thermometer
1 two-hole stopper
1 right-angled glass tubing, short legs
1 30-cm rubber tubing
1 beaker for cold water bath
1 small test tube

Safety Precautions: Goggles required. Alcohol is flammable and considerable care should be exercised. Do not allow alcohol vapors to escape into the room. A hot water bath must be used.

Procedures: The water bath must be heated slowly, so that the students do not miss when a fraction starts boiling. Students need to decide when to begin collecting the first fraction. Discussions should be directed toward increasing purity, e.g., re-distilling the fraction. Alcohol vapors need to be condensed in the test tube chilled by cold water.

*****

Experiment 8–3. Paper Chromatography

Objective: To develop another test for characteristic properties.

Problem Statement: Can paper chromatography be used to determine characteristic properties of a substance?

Supplies:

2 strips of filter paper
2 large test tubes

black ink
food coloring

Safety Precautions: Goggles required.

Procedures: The ink spot must be placed on the filter paper about 1 ½ centimeters from one end. The paper should be positioned in the test tube just above the water, without touching the water. Let the paper set for 5–10 minutes and observe.

****

Chapter 9

Experiment 9–1. The Properties of Some Common Gases

Objective: To distinguish between various gases.

Problem Statement: Can we identify a gas on the basis of its characteristic properties?

Supplies:
1 large test tube
1 large test tube clamp
1 pegboard support
5 small test tubes
1 rubber tubing, 30 cm
1 #4 one-hole stopper
1 test tube rack
magnesium ribbon or turnings (not powder)
1 Alka-Seltzer tablet
dilute sulfuric acid
3% hydrogen peroxide solution
limewater
natural gas jets

Safety Precautions: Goggles required. All of the precautions for working with acids and peroxides should be followed. Care should be taken to keep the room well-ventilated and to prevent the release hydrogen and natural gas into the atmosphere of the room. This will minimize the possibility of an explosive condition developing.

Procedures: Gas *a* is carbon dioxide and can be generated from Alka-Seltzer and water. Gas *b* is generated by warming a 3% hydrogen peroxide solution in a test tube. Students should be cautioned to just gently warm the generating tube. An alternative approach is to place manganese dioxide in a generating flask fitted with a two-hole stopper with a delivery tube and a funnel tube for pouring a 6% solution of hydrogen peroxide into the manganese dioxide. These approaches should be tested by the instructor prior to classroom use. Gas *c* is generated by placing six, 6.0-cm strips of magnesium ribbon into the large test-tube generator, adding 20 ml of 0.1-N sulfuric acid solution and quickly stoppering it with the delivery tube and stopper. Gas *d* is natural gas, which can be taken from the laboratory spigots.

The limewater solution may be prepared well in advance, and should be prepared at least one day before its use. A fresh supply should be prepared each semester. It can be made by dissolving one gram of calcium hydroxide in 50 ml of water, stirring vigorously and filtering. The clear filtrate is used for the carbon dioxide test. The limewater test should be done after the test tube samples are collected, especially in the case of the generation of carbon dioxide, as the reaction takes place so rapidly that most of the gas will be lost in the limewater test if it is done first. Bubbling the carbon dioxide through limewater generates calcium carbonate, which, because it is insoluble, causes the milky color. If too much carbon dioxide is bubbled through, however, the limewater clears again because the calcium carbonate becomes soluble.

*****

Experiment 9–2. The Decomposition of Water

Objective: To break water into its constituent gases.

Problem Statement: Can water be broken down into its components by electrical means?

Supplies:
2 small test tubes
2 solid stoppers
2 electrolysis electrodes
1 beaker, 250 ml or larger
1 6- to 12-volt DC power supply
2 leads
1 test tube rack
1 plastic spoon
sodium carbonate
wood splints

Safety precautions: Goggles required. If battery chargers or other power sources connected to a 115-volt supply are used, the standard safety precautions are to be observed. Be sure the electrodes are connected in the beaker, so they cannot short circuit one another. Adding the sodium carbonate and stirring must be done with a plastic spoon.

Procedures: Air bubbles may be eliminated from the test tubes by filling them with water until the water crowns up. The mouths of the tubes can then be covered with pieces of paper toweling, which seal onto the test tubes when they become wet. The test tubes can then be inverted into a beaker, half filled with water. The paper towel pieces can be flicked off in the water and retrieved.

During the experiment, the clamps should be spaced to eliminate any possibility of a short. The tip of the electrodes should protrude slightly into the mouth of the test tubes, so no gas is lost, but not so far that the tubes will not be able to fill completely with gas.

A six-volt lantern battery may be used for the energy source. A variable-voltage power supply will give students a feel for the effect of voltage on the speed of the reaction. After a minute of operation, two teaspoons of sodium carbonate should be stirred in and its effect on gas generation observed. The current should be stopped as soon as one test tube is filled with a gas. Both tubes should then be stoppered and placed upright in a rack.

*****

Experiment 9–3. Some Metals and Acids

Objective: To relate the relative reactivity of metals and acids.

Problem Statement: Are all metals affected the same by certain acids?

Supplies:
4 small test tubes
dilute hydrochloric acid
dilute nitric acid
dilute sulfuric acid
zinc, mossy
copper turnings
magnesium turnings

Safety Precautions: Goggles required. Follow all precautions for working with acids.

Procedures: Follow those given in the text.

*****

Experiment 9–4. Copper Oxide Reduction

Objective: To free metals from oxides.

Problem Statement: Can oxides be reduced to pure metals?

Supplies:
2 small test tubes
1 one-hole stopper
1 delivery tube, rubber
1 alcohol burner
wood splints
copper oxide
charcoal, powdered

Safety Precautions: Goggles required. Observe precautions working with heating test tubes. Be certain to remove the delivery tube from the water prior to removing the flame from the test tube so that cold water is not forced into the hot test tube, shattering it.

Procedures: Follow those given in text.

*****

Experiment 9–5. Interesting Juices

Objective: To distinguish acids from bases and salts.

Problem Statement: What are acids, bases and salts?

Supplies:
A variety of common acids, bases and salts
litmus paper
pH paper and an interpretation chart
soda straws
bromothymol blue
distilled water
microscope slides
250 ml beakers

Safety Procedures: Goggles required. Follow accepted laboratory safety procedures for working with acids.

Procedures: Follow those given in the text.

*****

Experiment 9–6. Draino and an Acid

Objective: To neutralize an acid with a base to produce a salt.

Problem Statement: Can we produce a salt by neutralizing an acid with a base?

Supplies:
2 medicine droppers
1 glass stirring rod
1 evaporating dish
phenolphthalein indicator
1 alcohol burner
1 burner stand

Safety Precautions: Goggles required. Draino is a strong base which dissolves organic material. If contact is made with the skin, it needs to be washed off with huge amounts of running water to get rid of the slippery feeling. Standard precautions for the hydrochloric acid should be followed. Dilute solutions of both should be used. The teacher can use experimentation to adjust the strength of each to be the same (0.1-N solutions recommended). Laboratory sodium hydroxide can be used as Draino. Materials should not be tasted in the laboratory although the temptation is strong to have the students identify the common salt, NaCl.

Procedures: Follow those given in the text.

*****

Problems

1. Increasing atomic number
3. 32
5. (a) 5; (b) 5; (c) 6; (d) 0
7. (a) A burning splint will cause it to burn. (b) A glowing splint will not be extinguished. (c) Invert a test tube of gas over an empty test tube and test it to see if the gas moves down, indicating it is more dense than air. (d) Bubble it through lime water.
9. Turn litmus blue; feel slippery; dissolve organic materials such as hair, wool, etc.; neutralize acids to form a salt and water; have pH greater than seven.

****

Chapter 10

Experiment 10–1. Wave Motion

Objectives:

1. To identify reflection of a wave from a fixed barrier.
2. To identify reflection of a wave traveling from a light medium to a heavy medium.
3. To identify reflection of a wave traveling from a heavy medium to a light medium.
4. To identify a standing wave.

Supplies: A slinky can be used for the large-diameter spring. The small diameter spring can be one similar to that used in a window shade roller. A piece of string can represent the light medium.

Safety Precautions: Goggles recommended. Springs may be easily overstretched. Discourage students from performing tricks with the slinkies such as having them move down stairsteps, as the focus should be on wave motion.

Procedures: Follow those given in the text.

*****

Experiment 10–2. Vibrating Strings

Objective: To identify the factors which affect the pitch of a vibrating string.

Problem Statement: In how many ways can you change the pitch of a vibrating string?

Supplies:

1 standard 30-cm ruler
a pencil
two wide rubber bands
1 tuning fork
1 beaker, 250 ml or larger

Procedures: Follow those in the text.

****

Chapter 11

Experiment 11–1. Reflection from a Plane Mirror

Objectives: To state the laws of reflection.

Problem Statement: How can the image formed in a plane mirror be explained?

Supplies:
1 3" x 3" mirror
2 common nails, longer than the height of the mirror
straight pins
protractor

Safety Precautions: Normal laboratory precautions should be followed.

Procedures: Students should understand that the light is reflected from the silver surface on the back of the plane mirror. A straight pin serves as an object and the sighting pins can be aligned to indicate the path of light. It should be pointed out that the virtual image only exists in the mind, as one is tricked into thinking that the light comes from behind the mirror. The image should also be located by parallax, as discussed in the chapter.

*****

Experiment 11–2. Images of a Concave Mirror

Objective: To state the conditions of the image formed by a concave mirror with the object located at the six significant locations.

Problem Statement: What types of images can be formed by a concave mirror?

Supplies:
1 concave mirror
1 battery
1 flashlight-sized bulb and socket
2 leads of appropriate length (30 cm)
modeling clay
2 meters of masking tape

Procedures: The first step should be to locate the focal point of the mirror. A piece of masking tape on the table can be the principal axis and allow marks to designate identified points such as *p*. Students should learn to reconstruct the images. It should be pointed out that light rays are drawn as a matter of convenience to indicate light paths, but that light does not travel as "rays." Wherever two rays coming from the same source intersect, a real image is formed. Two "imagined" rays form a virtual image if they intersect (as behind a plane mirror). Caution: Dry cells are designed to operate intermittently and should not be connected to a circuit for long periods of time.

*****

Experiment 11–3. Refraction of Light

Objective: To determine the index of refraction of a material using geometry.

Problem Statement: How can a general statement for the refraction of light be developed?

Safety Precautions: Goggles must be worn if liquids other than water are used. Follow normal laboratory safety procedures.

Supplies:
- A semi-circular, plastic sandwich box
- straight pins
- protractors
- various liquids that <u>do not dissolve the plastic dish and that are not a hazard to the students.</u>

Procedures: Various liquids can be checked, if time permits, to see how their refractive index compares to that of water. Trigonometry can be used to determine n, but the comparison of the chords helps to develop a conceptual model, without getting bogged down in mathematics.

*****

Experiment 11–4. The Converging Lens

Objective: To state the conditions of the image formed by a convex lens with the object located at the six significant locations.

Problem Statement: What types of images can be formed by a convex lens?

Supplies:
- 1 convex lens
- 1 battery
- 1 flashlight-sad bulb and socket
- 2 leads of appropriate length ($>$30 cm)

modeling clay
3 meters masking tape

Procedures: The converging, or convex lens, should be set up in a manner similar to the concave mirror. The differences in this case are that the light travels through the lens, instead of being reflected, as with the concave mirror. A candle flame can be used for the object, but an electric cell and light bulb are preferred.

*****

Experiment 11–5. Flame Tests

Objective: To determine which ions impart color to a flame.

Problem Statement: Is the color imparted to a flame a characteristic of an ion contained in the salt?

Supplies:
calcium chloride
copper chloride
copper nitrate
lithium chloride
potassium chloride
potassium nitrate
sodium chloride
sodium carbonate
strontium chloride
strontium nitrate
10 9-cm nichrome wires
10 balsa sticks, 6 mm x 6 mm x 10 cm
1 alcohol burner

Safety Precautions: Goggles required.

Procedures: Alcohol burners may be used as a heat source. The colors are more easily observed if the room is semi-darkened and safety should be stressed, as a flame and hot nichrome wires are used. A spectra kit may be purchased from science supply houses or one can be made by placing 5–10 g of various salts in vials. They can be supported by placing in holes that have been bored into a wooden block. Wire loops may be constructed by making small loops in the end of a 9-cm nichrome wire pushed into balsa wood handles. Holes for the handles can be bored into the wood support base and numbered 1–10, corresponding to the respective vials. Each loop should be kept with its respective vial throughout the experiment, so the chemicals do not become contaminated with one another. The wire loops should be heated in the bottom, outside edge of the flame, so that chemicals do not fall onto the wick of the burner, contaminating it with multiple colors.

*****

Experiment 11–6. Diffraction of Light

Diffraction of light kits, such as those referenced in the text may be purchased from science supply houses.

*****

Experiment 11–7. Emission Spectra of Elements

Objective: To identify elements by using a diffraction grating.

Problem Statement: Are the spectral lines of elements characteristic properties?

Supplies:
Plastic diffraction gratings mounted in 2 x 2 photographic slides
1 115-volt light socket
1 60-watt light bulb
1 showcase lamp
1 mercury vapor light source
an assortment of gas-discharge tubes
1 high-voltage source for gas-discharge tubes

Safety Precautions: Students must observe electrical hazard precautions when working around high-voltage sources used to activate gas-discharge tubes.

Procedures: If when observing a white light source, the spectrum is observed vertically, the grating should be rotated so the spectrum appears on both sides for appropriate viewing.

*****

Problems:

1. Inverted, real, larger than the object
3. Same size as the object, inverted, real
5. N = 0.71
7. Real, inverted, same size as object

****

Chapter 12

Experiment 12–1. Static Charges

Objective: To describe the law of electrical charges.

Problem Statement: Do electrical charges have characteristic properties?

Supplies:
1 vinyl strip
1 acetate strip
1 piece wool cloth
1 piece cotton cloth
1 pegboard support
2 test tube clamps
masking tape

Safety Precautions: Use standard laboratory precautions.

Procedures: Follow those described in the text. Electrostatic charges can be studied using materials available from science supply houses. Vinyl, acetate, wool cloth and cotton cloth can be secured locally.

*****

Experiment 12–2. Charging by Conduction vs. Induction

Objective: To charge a body by conduction and by induction.

Supplies:
1 electroscope (soda-straw type)
1 vinyl strip
1 acetate strip
1 piece wool cloth
1 piece cotton cloth
Masking tape

Safety Precautions: Follow normal laboratory safety procedures.

Procedures: Any available electroscope will work, including setting up your own by passing an electrode through a solid rubber stopper in a jar and stringing two pieces of thin aluminum foil on the end of the wire.

*****

Experiment 12–3. Series Light bulbs

Objective: To recognize a series electrical circuit.

Problem Statement: Do series circuits have characteristic properties?

Safety precautions: Follow normal laboratory procedures.

Procedures: Caution: Dry cells are designed for intermittent operation. The students should fasten only one lead to the battery (a collection of cells) and should leave the opposite lead free to momentarily touch the opposite terminal of the battery, so as not to drain the cells. Size D dry cells or lantern batteries may be used for the energy source. It is important that the light bulbs are all of the same size. Every student should experience unscrewing bulbs and refastening them sequentially. The intensity of the bulbs should be noted.

*****

Experiment 12–4. Parallel Light Bulbs

Objective: To recognize a parallel electrical circuit.

Problem Statement: Do parallel circuits have characteristic properties?

Supplies: The same materials used for series light bulbs with the addition of two more leads.

Safety Precautions: Follow normal laboratory procedures.

Procedures: Again, each student should experience the effect of disconnecting and reconnecting individual bulbs. The intensity of each bulb should be noted as bulbs are manipulated.

Caution: Dry cells are designed for intermittent operation. The students should fasten only one lead to the battery and should leave the opposite lead free to momentarily touch the opposite terminal of the battery, so as not to drain the cells. Size D dry cells or lantern batteries may be used for the energy source. It is important that the light bulbs are all the same size.

*****

Problems

1. Approach the electroscope without touching it. Ground the electroscope and remove the ground and then the rod.

3. (a) 30 Ohms; (b) 2 amps; (c) 5 Ohms: 10V, 10 Ohms: 20 V, 15 Ohms: 30 V, V_T = 60 V
5. R_T = 2.72 Ohms; I_t = 18.33 amps
   (a) 10 amps
   (b) 5 amps
   (c) 3.33 amps
   15 Ohms: V = 50 volts
   10 Ohms; P = 50 x 5 = 250 watts
7. $\frac{\text{charge}}{\text{sec}}$; ampere
9. Two conductors separated by a diaelectric; a) curent flows; b) discharge.
11. In series
13. Series/Parallel
15. Zero resistance

****

Chapter 13

Experiment 13–1. The Law of Magnets

Objective: To relate how the polarity of a bar magnet can be established.

Problem Statement: How can you establish the polarity of a bar magnet?

Supplies: 1 bar magnet, preferably without the polarity marked
1 piece of string long enough to support magnet from any overhead point

Safety Precautions: Follow normal laboratory safety procedures.

Procedures: This is a straight-forward experiment. Suspend the magnet in an area isolated from ferromagnetic materials. If a piece of magnetite is available, it would be a good idea to have the students manipulate it and some magnetic materials such as straight pins.

*****

Experiment 13–2. The Magnetic Field Around a Wire

Objective: To discover that a moving charge through a conductor carries an associated magnetic field.

Problem Statement: Will an electric current flowing through a wire affect

Problem Statement: Will an electric current flowing through a wire affect a magnetic compass?

Supplies: A size-*D* cell
a piece of insulated wire about 20 cm long
1 inexpensive magnetic compass

Safety Precautions: Normal laboratory safety precautions.
Procedures: Follow those given in the text.

*****

Experiment 13–3. The Electromagnet

Objective: To demonstrate the factors that affect an electromagnet.

Problem Statement: What factors affect the strength of an electromagnet?

Supplies: Insulated wire
1 bolt
battery

Safety Precautions: Follow normal laboratory safety procedures.

Procedures: The students should observe the magnetic strength by varying the number of cells holding the number of turnings constant as well as by holding the number of cells constant by varying the number of turnings.

*****

Experiment 13–4. Electromagnetic Induction

Objective: To relate the relationship between a moving magnetic field and electrical current generation.

Supplies: About 2 m of insulated bell wire
1 bar magnet
1 galvanometer

Safety Precautions: Follow normal laboratory safety procedures.

Procedures: This experiment can be done as a demonstration, if you have a limited number of galvanometers.

Index

A

B

C

D

E

H

I

J

K

L

M

Q

R

S

T

U

V

W

X